The Moral Dimensions
of Politics

The
Moral Dimensions
of Politics

RICHARD J. REGAN, S.J.

New York   Oxford
OXFORD UNIVERSITY PRESS
1986

Oxford University Press

Oxford   New York   Toronto
Delhi   Bombay   Calcutta   Madras   Karachi
Petaling Jaya   Singapore   Hong Kong   Tokyo
Nairobi   Dar es Salaam   Cape Town
Melbourne   Auckland

and associated companies in
Beirut   Berlin   Ibadan   Nicosia

Library of Congress Cataloging-in-Publication Data
Regan, Richard J.
The moral dimensions of politics.
Bibliography: p.   Includes index.
1. Political ethics. I. Title.
JA79.R45   1986   172   85-21807
ISBN 0-19-503974-2
ISBN 0-19-503975-0 (pbk.)

Printing (last digit): 9 8 7 6 5 4 3 2 1

Printed in the United States of America
on acid-free paper

Preface

One of the occupational trials of college teachers is to select suitable texts for courses. Over the years—twenty to be exact—I have found this task particularly difficult for courses dealing with the moral foundations of politics and the relation of moral principles to public policy. Despite the existence of many excellent books and articles on one or another aspect, there is no single book which presents a broad overview of morals and politics from an Aristotelian-Thomistic perspective and relates that perspective to public policy. The purpose of this book is to fill the lacuna. I begin at the beginning with chapters on the nature of morals and politics, develop general principles of political justice, and apply Aristotelian-Thomistic principles to three major policy areas: public morals, social justice, war and peace.

I make no claim that my explanation of Aristotelian-Thomistic principles is original, and I do not claim that inferences I draw from those principles for public policy are other than my own. The purpose of the book is to make available to students and the general public a primer on Aristotelian-Thomistic principles and their implications for major public policy areas. I hope that the wisdom of that tradition of reason will enable readers to weigh issues of public policy and to understand the philosophical framework within which many, like the American Catholic bishops, formulate statements on abortion, economic justice, and nuclear war.

The book will be useful for college courses in political philosophy and adult discussion groups. The first three chapters could constitute a nucleus of material for courses on the purposes of politics, Chapters 2, 3, and 5 a nucleus for courses on social justice, and Chapters 2, 6, 7, and 8 a nucleus for courses on the morality of war. The bibliographical note suggests selective readings from primary and secondary sources to complement the text.

I have concentrated on principles rather than cases. Teachers may wish to supplement the text with case studies or analyses of specific public policies. On public morals, the case of abortion could be considered in greater detail, and other cases of public morals added. On social justice, cases of business ethics and corporate responsibility at the national and international level could be considered. On war, the morality of past and present U.S. military activities in foreign countries could be considered. Students generally find case studies more attractive than general principles, and teachers may wish to treat more cases in class or to assign case studies to students as term papers.

In view of my general purpose, I have tried to chart a course between the Scylla of too little scholarship and the Charybdis of too much. My fellow teachers and their students will have to decide how far I have succeeded in doing so. The book touches a large number of topics and issues, and scholars will undoubtedly seek fuller treatment than I provide in the text. I hope that footnotes and the bibliographical note will suggest leads for further study. Introductory students, on the other hand, may find the arguments in the text rather abstract and the illustrations too few or unfamiliar. To ameliorate this difficulty, I rely on the talent of teachers to explain the text with the aid of supplementary readings.

I have adapted a few pages of material from previously published books of mine, *American Pluralism and the Catholic Conscience* (New York: Macmillan, 1963) and *Private Conscience and Public Law: The American Experience* (New York: Fordham University Press, 1972), and from several articles of mine in *Thought* magazine. The author gratefully acknowledges permission of the publishers to use the material here.

The Rev. Francis P. Canavan, S.J., the Rev. Joseph C. McKenna, S.J., and the Rev. Andrew C. Varga, S.J., have been particularly generous with their time in reading the manuscript and offering helpful suggestions. To these and to all who assisted me in many ways, I offer my sincere thanks.

Bronx, N.Y.   R. J. R.
January 1986

Contents

Sigla

CE   Thomas Aquinas, *In libros Ethicorum expositio* [*Commentary on the Ethics*]

CP   Thomas Aquinas, *In libros Politicorum expositio* [*Commentary on the Politics*]

CS   Thomas Aquinas, *Scriptum super libros Sententiarum Petri Lombardi* [*Commentary on the Sentences of Peter Lombard*]

DR   Thomas Aquinas, *De regno, ad regem Cypri* [*On Kingship*]

JL   John Locke, *The Works of John Locke*, 12th ed., 9 vols. (London: Rivington, 1824)

LN   Thomas Hobbes, *Leviathan*

NE   Aristotle, *Nicomachean Ethics*

ST   Thomas Aquinas, *Summa theologica* [*Summary of Theology*]

TG   John Locke, *Second Treatise of Government*

The Moral Dimensions
of Politics

1

The Human Person and Moral Norms

The thesis of this book, to put the matter simply, is that political good is essentially linked to moral good. Before we can consider that linkage, however, it will be necessary to examine the nature of moral good itself. The pluralist character of modernity may seem to situate the quest for commonly accepted moral norms beyond our reach. In the field of ethics, the most bewildering variety of points of view exists. Indeed, a large percentage of our contemporaries would even reject any notion of ethics and interpret all human behavior in terms of subjectively motivated self-interest.

We can begin by considering the so-called "ethical fact." All of us make judgments about what should or should not be done—frequently with greater clarity and intensity about the behavior of others than about our own. Even dedicated egoists do. In this sense, an "ought" represents an "is." The nature of the problem, then, is to discover the objective grounds for ethical judgments so that we can distinguish correct judgments from those that are faulty.

This chapter will describe and defend the Thomistic natural theory of morals. This will involve explanation of key concepts drawn from Greek philosophy: "nature," "reason," "freedom," and "good." We then look at how Aquinas synthesized the Greek concept of moral good and the biblical concept of moral law to articulate the theory of natural law. We examine the distinction between primary and secondary precepts of the natural law and the distinction between secondary precepts easily accessible to reason and those which require greater reflection, with the precepts against lying and killing serving as illustratory examples. We especially consider the question of whether, or to what extent, precepts of the natural law are universally valid. Lastly, we describe how conscience relates the objective order of morality to the subjective order of individual responsibility.

There are, of course, other ways of systematizing morals, and we cannot do justice to any or all of those here. We can, however, point up how two prinicipal theories, Utilitarianism and Marxism, and their Christian derivatives differ fundamentally from Thomistic theory.[1] The central Utilitarian principle, "the greatest good of the greatest number," has been interpreted by adherents in many different ways. Some Utilitarians interpret the principle to mean that the norm of morality is the maximum amount of pleasure accruing to members of society irrespective of how many or how few individuals derive pleasure from human acts, while others interpret the principle to mean that the norm of morality is the maximum distributional spread of pleasure among members of society. Both of these interpretations are known as *Act Utilitarianism*. *Rule Utilitarianism*, on the other hand, interprets the principle to justify adherence to a system of rules unless a calculus of maximum pleasure or utility in the face of conflicting pleasures or utilities concludes to the contrary. All of these versions of Utilitarianism have in common the basic conviction that the morality of human action is to be judged solely by results or consequences.

The same consequentialist assumption lies at the root of Marxism, although Marxists have no summary principle like the Utilitarian "greatest good of the greatest number." Since Marx held the view that the historical process inevitably determines events, ethics becomes simply a matter of whether or not human activity furthers the process. The ultimate end of the process, after the state withers away, is the perfect reconciliation and rationalization of self-interests in comunitarian interests. The intermediate steps toward that end are the heightening of proletariat class consciousness, revolution, and a dictatorship of the proletariat to reconstruct society and social consciousness. From this perspective, the moral test of human activity is only whether the net result of the activity contributes to stages of the historical evolution toward the final end. It is of no matter that ten million socially unreconstructed kulaks are liquidated, or that individual freedoms are denied, so long as these actions contribute to the historical process. In other words, the morality of human action is to be judged solely by its consequences for the historical process.

Utilitarianism and Marxism are purely secular ethical systems, but they have their respective religiously connected imitations. Situation ethics is a Christian imitation of Utilitarianism. According to Joseph Fletcher,[2] a foremost spokesman for and popularizer of situation ethics, Christian ethics must avoid two extremes: legalism (rules determine morality) and antinomianism (morality has no rules). Love offers a middle course between these extremes, and love is the only standard of action in particular situations. This may sound quite different from secular Utilitarianism, but is it? True, where Utilitarians calculate the greatest good of the greatest number in terms of pleasure or self-interest, situation ethicists calculate the greatest good of the greatest number in terms of neighbor-regarding love. But Fletcher's examples make abundantly clear that the "love" at stake in moral decisions is not directed toward particular individuals and their welfare. Rather, one coldly calculates the greatest amount of neighbor welfare for the greatest number of neighbors. In this essential respect, situation ethics is as consequentialist as Utilitarianism.

The same fundamental criticism can be made of Christian imitations of Marxism, of which liberation theology is a popular contemporary example with a wide following among priests and intellectuals in Latin America and elsewhere.[3] Liberation theologians condemn capitalist society root and branch and, according to their critics, support the use of indisciminate revolutionary violence to overthrow existing capitalist structures. Since individuals participating in capitalist structures thereby contribute to the institutional violence that those structures inflict on the poor, counterviolence without regard for the individual responsibility of targets is fully justified. Thus, liberation theologians, according to critics, do not view love primarily as a relationship between individuals but rather as regard for the liberating consequences for underprivileged masses of people. Liberation theologians, in the critics' view, appear to regard particular individuals as dispensable without respect to their personal guilt or degree of involvement with capitalist structures. Like their unbaptized Marxist counterparts, liberation theologians seem to make the end justify all means.

Mention should also be made of one other alternative to natural law based on religious premises. Protestant Reformers of the sixteenth century thought the fall of Adam from original grace so total that human nature as a result became radically depraved, and all human acts inherently sinful, even with justification in Christ. As applied to ethics, this theological position can be summarized in what has been styled the "Protestant Principle," namely, that human will is too debased to act in unambiguously moral ways, and human reason is too enfeebled to recognize moral norms clearly without the aid of divine revelation. It is inappropriate to analyze in this place theological aspects of Protestant belief, but it is appropriate to ask whether or not the Protestant view of sinful man needs to exclude the concept of natural law entirely. As the Protestant theologian, Dr. John C. Bennett, has observed, "surely there is a moral order which human laws do not create. . . . There are differences here [among religious communities], but they should not obscure what is common to our traditions in contrast to secular moral relativism. . . . I do believe that the objective moral order does press upon us even when we do not recognize it."[4] Without some concept of a natural moral order, Christians risk having nothing of relevance to say in moral matters to the millions of people who are not Christians. To put the point in the form of a question, how can Christians take positions on the moral management of society without appealing to reason?

Greek Moral Good

Aristotle is a systematic representative of Greek moral thought at its zenith in Athens of the fourth century B.C.. In line with an overall teleological perspective,[5] Aristotle viewed humans as intrinsically ordered to a specifically human good. This human good, which brings us to a state of well-being and so makes us truly happy, consists in activities of reason itself and other activities in accord with right reason.[6] Moral virtues, such as those of courage, moderation, and justice, characteristically dispose us toward good action, and the architechtonic intellectual virtue of moral reasoning, prudence or practical wisdom, determines the appropriateness of means to achieve moral virtue. Lastly, and most important, Aristotle situated the quest for virtue quintessentially within a structured society with other humans. For Aristotle and ancient Greeks generally, only those who were more than human (pure spirits or superhuman) or less than human (brutes or quite literally "uncivilized") would be happy apart from the *polis*, the Greek city-state. Hence, there could be no genuinely good or humanly happy person apart from the *polis*, and Aristotle accordingly devoted the whole of the *Politics* to the right ordering of the *polis*, that is, to an ordering of the *polis* that would promote morally virtuous citizens.

Aristotelian epistemology is altogether realist: reality is the measure of consciousness, and consciousness is a window on reality. The capacity of the human mind to know the real is evident to each individual in the actual experience of the real. In that transparency of experience, humans know that their consciousness opens on a real world of material objects and human subjects, and that their consciousness absorbs and responds to the real rather than constructs or posits its content. Moreover, and even more important, the human intellect is capable of understanding the structure of reality.

The ability of human reason to understand the nature of things needs to be carefully distinguished from other functions of intellect which we moderns more easily recognize. We moderns are accustomed to equate human intellect or reason with "reasoning," that is, with the movement of the mind from premises to conclusions. We "reason" in this way both speculatively and practically. In empirical sciences, scientists employ a hypothetical-deductive method of reasoning. Scientists first hypothesize a theory to explain observed events, deduce consequences from the theory that can be observed, and then test to see if those deduced consequences do occur. In mathematical sciences, mathematicians employ pure deduction to draw conclusions from postulated premises. In everyday affairs, we employ both inductive and deductive reasoning to reach conclusions about everything from tomorrow's weather to choosing the best route in rush-hour traffic.

Aristotle identified an additional function of the human intellect, and this is the function with which we are principally concerned here: the capacity of the human intellect to understand and form concepts about the basic structure of things. In the case of human beings themselves, we are capable of understanding from our experience that we engage in rational activities not only in the sense of reasoning about things from premises but also in the sense of understanding the structure and purpose of things in their own right. Of course, one can reason from premises to conclusions about the nature of things. For example, one may conclude that humans have the capacity to use speech to convey thought and to engage in humor precisely because humans are rational. But the function of the human intellect to understand the nature of things without reasoning from premises is at the heart of Aristotelian philosophy generally and Aristotelian ethical theory specifically. This is the function of reason which is absent from much modern philosophical thinking and inappropriate for modern science.

The central theme of Aristotelian metaphysics is that the structure of things determines their type of activity and, conversely, that the specific type of activity indicates the structure of being. Activity develops and perfects beings according to their species, that is, brings them to their specific ends and perfection. Trees, horses, and human beings engage in different types of activities, and each type of activity brings the respective species to its perfection: seedling to full-grown tree, colt to fully developed horse, baby to adulthood. Reason is able to understand from these data of experience that activity is purposeful, ordered to ends. "Nature" simply denotes the structure of being when considered as the immanent principle of the activity, and "human nature," the common structure which members of the human family share when that structure is considered as the ultimate principle of activity directed toward human perfection.

The goal of human activity, then, is the well-being proper to the human species, and Aristotelian psychology recognizes rationality in all its dimensions as the characteristic that distinguishes human activity from that of other species. We humans, of course, have animal appetites which we act to satisfy: we eat to satisfy our hunger, we seek to avoid pain, and we may engage in sexual activity. But as long as we act consciously, our animal activities are *also* rational. That is to say, we have to make rational decisions about our animal activities: we have to decide when and how much to eat, how to avoid and alleviate pain, and when and with whom to have sex. We rationally perceive food and sex as good for us and pain as bad for us—but only as partially good or bad—and so we are presented with choices in our activities. (Aristotle himself is less than luminous or complete in his explanation of human freedom, but he nonetheless clearly affirms its existence.) Humans can understand the basic structure of themselves and their world, but only by their free acts do humans determine how they are live in their world. Moreover, Aristotle insisted that human intelligence and freedom open us not only to the physical universe but also and especially to a life of human intersubjectivity—to the friendship and love of other human beings. The latter aspect of Aristotelian psychology needs to be emphasized here and will be elaborated in Chapter 2: humans are not autonomous individuals but oriented by nature to community.

Human well-being consists in activities of reason, intellectually virtuous activities, and other activities, according to judgment of right reason, morally virtuous activities. If humans so act, they achieve their human well-being and happiness. We should note that Aristotle's concept of moral obligation is purely hypothetical: if humans wish to achieve their specific goal of human well-being, they should engage in activities of reason and other activities according to right reason. We cannot choose any ultimate good except that specified by nature, but we are free to choose means conducive to, or destructive of, achievement of the ultimate good specified by nature. Virtue rewards humans with happiness because they have achieved their proper good, and vice punishes humans with unhappiness because they have not.

Aristotle's general norm for moral behavior is a mean between the extremes of too much and too little: don't eat too much or too little; don't act rashly or cowardly; so deal with others that they receive no more and no less than is their due. By observing this mean, humans develop capacities for morally good action into the characteristic dispositions constituting moral virtues. Aristotle relied on a rightly ordered *polis* to provide more specific norms of human behavior. In Book Two of the *Nicomachean Ethics* (2. 6. 1107^{a}10), in the course of denying any mean between virtue and vice, Aristotle gives three examples of universally vicious acts: murder, theft, and adultery. But we should note that he immediately adds that these acts are wrong by name or definition, presumably because he understands murder as unjustified homicide, theft as taking property belonging to another, and adultery as intercourse with a woman belonging to another.

Whether or to what extent Aristotle integrated natural moral good into a framework of moral law is a disputed question. Before examining specific Aristotelian texts relevant to a concept of natural law, it is useful to recall that Aristotle's "God," the *noesis noeseos*, unlike the Judeo-Christian God, did not create the world and exercised no analogously efficient causality on the preexisting matter which Greeks thought eternal. Thus, at the outset, we should be wary of claims that Aristotle held any developed concept of natural law.

The principal text adduced to support the claim that Aristotle held a version of natural law is contained in Book Five of the *Nicomachean Ethics*. Aristotle there (5. 7. 1134^{b}18–30) distinguished what is just by nature (*dikaion physikon*), which has the same validity everywhere, from what is just by convention (*dikaion nomikon*), which could be determined in other ways. Aristotle is expressly speaking of "political justice," that is, what is just in the *polis*. Accordingly, Aristotle may be distinguishing in this place what is just by nature, that is, what is just because it is integral to the right ordering of the *polis*, from what is just by arbitrary convention. The text may also be plausibly read to distinguish acts that the *polis* commands (e.g., restitution of stolen goods) or prohibits (e.g., murder) on the basis of nature from those acts that the *polis* commands (e. g., driving on the right side of the road) or prohibits (e.g., billboards alongside public highways) where nature has made no determination about right and wrong. In any case, while Aristotle undoubtedly refers to a natural *order*, it is not clear that he refers to a natural *law*.

In the *Rhetoric*, Aristotle seems to come closer to an explicit natural law position. In Chapters 13 through 15 of Book One, Aristotle distinguishes general from particular law, and general law is said (1373^{b}1-2) to be based on nature (*kata physin*) because humans have a common idea of what is just by nature (*dikaion physei*). Lawyers with a weak case in written law are urged to appeal to equity and the general, unwritten law, as the Greek playwright Sophocles' character, Antigone, did to justify her disobedience of King Creon's command not to bury the body of her brother, Polynices (1375^{a}6).[7] But the *Rhetoric* is offering advice to would-be lawyers, and there is no explicit assertion in this text of the *Rhetoric* that the general, unwritten law, however much based on nature, is a natural law (*nomos physikos*). In fact, Aristotle's description of the unwritten law as general (*koinos*) suggests a sort of *jus gentium* or customary law deriving its validity as law from its common acceptance and practice rather than simply from the natural law underlying it.

If Aristotle lacked a theory of natural law, or at least a developed theory, he nonetheless provided a solid basis for the subsequent development of such a theory. Aristotle recognized that human beings are intrinsically ordered to the specific good of rational activity and activity in accord with right reason. We can now turn to the biblical doctrine of creation that enabled Christian thinkers like Aquinas to reinterpret the Aristotelian natural order of moral good as a natural law imposing moral obligation.

Biblical Moral Law

The biblical Israelite moral perspective was quite different from that of Aristotle and Greek philosophy. What Yahweh commanded was morally obligatory, and what Yahweh prohibited was morally sinful. In the earliest pages of the first book of the Old

Testament, the writer of Genesis describes how "the man and his wife hid themselves from the presence of the Lord God among the trees of the garden" after eating the forbidden fruit (Gen. 3:8). Thus, the writer of Genesis attributes to Adam and Eve a sense of moral guilt subsequent to, and because of, their disobedience of Yahweh's explicit command. Later, the covenant between Yahweh and His people, however much a loving relationship, brought with it the obligations of the Torah and the Levitical law. It is true that the prophets summoned the people of Israel from merely external observance of the Law to interior worship of, and fidelity to, Yahweh. But the prophets did not thereby challenge the claims of the Mosaic law to obedience from the Jewish people.

At the time of Jesus, the Pharisees and other Jewish leaders are represented in the Gospels as more preoccupied with external observance of the Law than with interior dispositions in the achievement of religious virtue. As represented in the Gospels, Jesus attacked this preoccupation of the Pharisees with ritual observance by arguing that it is not what goes into a man's mouth which defiles him but what comes out therefrom, since "what comes out of the mouth proceeds from the heart" (Matt. 15:17–18). But the Gospels do not portray Jesus as abrogating the claims of the Law to obedience from the people, although the Gospels do portray Jesus in the tradition of the prophets on the primacy of interior dispositions over merely external observance, and as more liberal in interpreting the Law than His Pharisaic adversaries.

Christian evangelization of the Gentiles shifted the focus of moral concern beyond the ethnocentric and theocratic confines of the Jewish people. It was the Apostle Paul's lot to declare Gentile converts to Christianity freed from any obligation to observe external rituals of the Law. But in the same Letter to the Romans, which has as its main theme that Christians are freed by faith in Christ Jesus from obligations to observe external rituals of the Law, paradoxically enough, Paul holds all peoples bound to observe the Law's essentials. In a celebrated passage in the second chapter of Romans (2:14–16), Paul claims that the Gentiles have essentials of the Law written in their hearts, that they carry out these precepts by nature (*physei*), and that their consciences will condemn them on the day of judgment if they fail to do so. Thus, the essential demands of Yahweh on the people of Israel by positive command, as expressed in the Ten Commandments, are now seen as moral demands on all peoples by the constitution of human nature itself.

Thomistic Natural Law

Thomas Aquinas' moral philosophy is a landmark in the history of Western thought. Aquinas synthesized Greek moral good, as represented by Aristotle, and biblical moral law, as represented in the Ten Commandments, into a cohesive natural law theory. Roman interpreters of Stoic philosophy like Cicero, more legally oriented than the Greeks, did modify the Greek ideal of moral good by identifying human reason with natural law and moral consciousness with legal obligation. But whatever the contributions of Stoic philosophy to natural law theory, earlier Greek philosophies of natural moral good and the biblical tradition of divine moral law provided the two central components of the Thomistic synthesis. With these components in place, it was but a short step to interpret natural moral good in terms of moral obligation because natural moral good is part and parcel of God's creative act.

Thomistic natural law recognizes a provident God who creates purposefully. In doing so, God not only causes creatures to be, He causes them to be in the specific ways in which they exist and orders them to their specific goals. Aquinas, thus, integrated the Aristotelian philosophy of nature into the biblical theology of creation, contending that even the latter doctrine is accessible to human reason without the aid of divine revelation (ST 1. 44. 1).

Since the fact of creation is accessible to human reason, Aquinas is in a position to reinterpret natural moral good as natural moral law. Aquinas identifies four essential characteristics of law: (1) law rationally orders means to ends; (2) the end of law is the good of the community; (3) those with authority over and responsibility for the community enact law; (4) law is published to members of the community (ST 1-2. 90. 1-4). This definition, of course, derives from our experience with human law, but Aquinas shows that God's creative act also satisfies the definition analogously. By the very act of creation, God orders created things rationally toward specific goals and specific goals toward a general goal, obviously has authority over and responsibility for the community He creates, and manifests the ordinance in the natures He creates. Since the act of creation itself is identical with God Himself, that act can be properly called eternal law. Thus, Aquinas concludes that God as Creator can be rightly thought of as a legislator, and His creative act as an eternal law (ST 1-2. 91. 1).

Creatures other than humans have no freedom whether to act or not to act in ways directed toward their specific perfection. Humans too are not free to determine their specific goal or the relation of means to that goal. But because humans are free to act or not to act in ways conducive to their specific goal, humans do participate in the eternal law rationally and freely. By reflective intelligence, humans discern the structure of their being and the relation of prospective action as a means to achieve their proper perfection, and humans by their free choice act (or fail to act) to achieve their proper perfection. When humans act according to their nature, they share with their own reason and will in God's plan for themselves as individuals and as a community. This human participation in the eternal law can be properly called natural law (ST 1–2. 92. 2). Note that Aquinas, by assigning a participatory role to humans in legislating the eternal law for and to themselves, recognizes the necessity of personal autonomy for authentically moral decision making.

The Thomistic natural law, then, is the entire set of commands and prohibitions deriving from God's creative act but self-imposed by the personal use of human reason. These commands and prohibitions derive not from human will but from the structure of a total and integrally constituted human nature. Because humans are intelligent and free creatures, the eternal law for them is a *moral* law, that is, a law which they rationally discern and freely decide whether or not to ratify. If humans ratify the eternal law for themselves, they fulfill their nature, achieve their well-being, and so carry out God's plan; on the other hand, if they reject the eternal law, they deform their nature, fail to achieve their well-being, and so vitiate God's plan.

Before we consider how reason discovers the natural created order and its moral demands on humans, it is important to stress that Thomistic natural law theory is not a specifically theological doctrine but rather a philosophical explanation of the foundation of morals. It is true, of course, that Aquinas was a Christian theologian, and that he considered creation—and, therefore, natural law—as simply one aspect of God's salvific plan for mankind. Where Greeks, like Aristotle, rested content with human goodness in a self-sufficient *polis*, Aquinas was concerned as well with Christian holiness, with obedience to the commands of God explicit or implicit in His creative act and salvific will, and, conversely, with the sinfulness of human actions that vitiate His salvific will. In this sense, the natural law has a supernatural dimension: observance of the natural law is a grace-ful act, and failure to observe the natural law is a sin-ful act. It is also true that divine revelation, in Aquinas' scheme of things, plays a strong supportive role in recognizing demands of the natural law, and divine grace an indispensable role in observance of the natural law. Lastly, it is true that Aquinas regards Christian love as the guiding priniciple of all virtuous action.

But the Thomistic analysis and argument for natural law remains one accessible to human reason irrespective of the divine salvific design. Reason itself can discern an order in nature that makes moral demands on human behavior. Because these demands do derive from an order of nature, their observance rewards humans with the contentment of self-fulfillment, and their nonobservance punishes humans with the discontent of a failed nature. Nature entails this sanction independently of the supernatural dimensions of human acts in the Christian dispensation. From this perspective, we can say that Aquinas introduces God's creative act to provide an ontological ground for a natural ethic based on the goal of truly human happiness; he brings God into the act, so to speak, to explain the ground of a eudaimonistic or happiness ethic, not to supplant it. We should strive to be humanly happy not only because human happiness is our human goal, but also because human happiness is the goal specified by God.

Aquinas' theory of natural law will also be inadequately understood unless it is related to his treatment of moral virtue. Natural law tells us what we should do and what we should not do. Just how natural law "tells" us this we shall discuss in connection with the role of conscience. Here we note the consequences of individual acts for habitual behavior. The performance of morally good acts develops habits or characteristic dispositions toward morally good behavior, and the performance of morally bad acts develops habits or characteristic dispositions toward morally bad behavior. As a result of repeated good acts, we develop facilities for acting rightly in the future. Thus individual acts have a moral import beyond the individual act itself: good acts build up the capacity to act rightly. And so Aquinas is concerned not only with the moral goodness of individual human acts, but also and especially with the disposition toward good action that results from the individual acts.

Existentialist philosophers have made us quite conscious of the fact that there is more to human fulfillment than action conformed to nature. No doubt humans can adhere to the demands of the natural law and yet fail to satisfy higher and richer potentialities of the human person; friendship and love alone can supply a completeness to human existence. In the minds of many moderns, this existential dimension of human life leaves no place for the ethical essentialism of Thomistic natural law. Yet, although it is true that conforming human actions to human nature does not assure the existential fulfillment of friendship and love, it does not follow that existential fulfillment itself is unrelated to the human essence. What radically existentialist philosophers overlook is the reality that human existence is a structured existence, not a pure existence. Hence, even the richest activities of the human person, if they are to fulfill human personality itself, need to be proportioned to the objective structure of the human person. Friendship and love indeed offer the maximum opportunities for human activities to be fully human, but the participatorily self-imposed natural law sets the minimum standard for those activities to be truly human at all.

Primary and Secondary Precepts

According to Aquinas, the absolutely first principle of practical reason, for example, the absolutely first principle of human action, is that humans should seek what is humanly good for them and avoid what is humanly bad for them (ST 1-2. 94. 2.). This principle is self-evident: "good" is the perfection or goal that all beings seek to achieve by their activities, and humans, simply by understanding the terms involved, can assent to the proposition that humans should seek what is naturally good for them and should avoid what is naturally bad for them. As concepts, the predicate "what humans should seek" is so contained in the subject "human good," and the predicate "what humans should avoid" is so contained in the subject "human evil," that no reasoning process, inductive or deductive, is necessary before reason can assent to the connection between the subjects and predicates.

The absolutely first principle of practical reason is also self-evident in its accessibility to the human reason of all; all humans with the use of reason understand the terms of the principle and assent to the principle upon understanding the terms. And the principle is absolutely first in both the logical and the ontological orders. The principle is first in the logical order because no moral norm can be articulated without presupposing that moral good should be sought, and that moral evil should be avoided. The principle is first in the ontological order because it articulates the most general truth about moral obligation itself.

The fact that the absolutely first principle of practical reason is a self-evident proposition does not make it a tautology, a proposition whose predicate only restates its subject in other words, after the manner of a mathematical definition. The concept of moral obligation to seek human good is indeed virtually contained in the concept of human good, but only when we understand the relation of human good to human action do we also understand that humans should seek their specifically proper good. Since the absolutely first principle of practical reason that humans should seek what is humanly good and avoid what is humanly bad is a proposition to which the intellect assents without any inductive or deductive reasoning process, it is a truism, albeit a fundamental truism.

Because good is the goal of action, we also understand that those things toward which human nature inclines us are human goods, and that their contraries are human evils (ST 1-2. 94. 2). On the basis of experiential knowledge, Aquinas identifies three classes of objects toward which human nature inclines us. In the first place, nature inclines humans, in common with all substances, to strive to preserve their being. Secondly, nature inclines humans, in common with all animals, to seek sexual union and children. Thirdly, nature inclines all humans precisely as rational to seek after truth and companionship with other humans. Practical reason, on understanding these objects to be objects of inclinations from nature, articulates primary precepts of the natural law: humans should seek in reasonable ways—but only in reasonable ways—to preserve their lives, to enjoy sexual union, to procreate and educate their children, to seek truth and understanding, and to live cooperatively with others in society.[8] As in the case of the absolutely first principle of practical reason, humans can assent to the primary precepts of the natural law without any inductive or deductive reasoning process, and all humans with the use of reason do assent to the precepts on understanding the precepts' terms.

Reason, with knowledge gained from experience, is able to draw conclusions about particular types of human acts from the first, general principles, and these conclusions constitute secondary precepts of the natural law (ST 1-2. 94. 4). Aquinas admits that some secondary precepts may be so formulated as to be valid only for the most part. For example, one formulation of a secondary precept of the natural law prescribes that property held in custody for its owner should be returned to the owner on request, but it is obviously unreasonable to return a gun to an owner with homicidal intentions. Moreover, Aquinas admits that not all persons and peoples will recognize the validity of secondary precepts when they are valid. For example, a secondary precept of the natural law prescribes that one should not appropriate the property of another, but passions, bad habits, or customs may blind the reason of some persons and peoples from recognizing the precept. Thus, some secondary precepts may be so formulated as to be valid only for the most part, and some secondary precepts, whether so formulated as to be universally vaild or not, may be recognized only by most persons and peoples.

Aquinas explains why secondary precepts of the natural law can be so formulated as not to be universally valid by contrasting the conclusions of practical reason with those of speculative reason (ST 1-2. 94. 4). Speculative reason deals with necessary things, and so its conclusions as well as its first principles purport to be universally valid. But practical reason deals with contingent things, and contingencies vary widely. There is an essential difference between methods of speculative and practical reasoning: speculative reason draws conclusions from principles in a purely (mathematics and philosophy) or hypothetically (natural science) deductive way, while practical reason (ethics) draws conclusions by relating concrete experience to principles.

As important as the distinction between primary and secondary precepts of the natural law is the distinction within the class of secondary precepts between those which are proximate and those which are remote (ST 1-2. 100. 1). The relation of certain human acts to first principles is so easily understood that reason can promptly, that is, with little reflection, conclude to a general judgment that the acts are to be approved or disapproved. The relation of other human acts to first principles is not so easily understood, and so considerable reflection on contingencies is required before reason can conclude that the acts are to be approved or disapproved. The latter remote conclusions are the province of those Aquinas calls "wise men," and we call professionals. Aquinas does not elaborate on differences between the types of conclusions, but we can illustrate differences between the two from Aquinas' reasoning about the morality of particular human acts.[9]

Precepts of the Decalogue are concerned with the order of justice, and the second part of the Decalogue with what one human owes to another (ST 1-2. 100. 5). According to the first principles of practical reason, human evil is to be avoided, and humans should act reasonably toward the others with whom they are naturally inclined to live. From these principles, one can conclude with little reflection to general norms to honor parents, not to kill, not to commit adultery, not to steal, not to lie, and not to covet one's neighbor's wife or goods. These precepts of the Decalogue, and indeed all precepts of the Decalogue,[10] are proximate precepts of the natural law (ST 1–2. 100. 1, 3, and 11). Special divine promulgation of the Decalogue, nonetheless, may be helpful because proximate secondary precepts of the natural law are not self-evident and require reasoning, that is, they might not be recognized by some without divine promulgation.

Examples: Precepts Against Lying and Killing

Considerable reflection may be required before we can apply the second part of the Decalogue to a variety of concrete situations. In the first place, it is necessary to understand the nature and purpose of human activity, and how a prospective act would vitiate that nature and purpose. The conclusion that lying is wrong, for example, rests on a clear understanding of the communicative purpose of human speech, and that lying is incompatible with that purpose. The principal purpose of human speech is to communicate the state of one's mind to another, although speech can be used to tell jokes or to act out roles on the stage. When one does purport to represent one's mind to another and does so falsely, this runs counter to the communicative purpose of speech. Put another way, truthful communication is integral to living reasonably in society with other human beings, toward which we are inclined by nature, and lying is basically antisocial. Since God's creative act ordered human speech to this social end, and reason can recognize this order, it is a precept of the natural law not to lie.

But reflection on experience also leads reason to understand that humans are not always under an obligation to communicate their thoughts, and that humans may even be under an obligation not to communicate their thoughts. We may use language in socially acceptable ways so as not to communicate our thoughts. Language communicates meaning only in a social context, and so the very same words can communicate different meanings in different contexts. When told by a seller that a used car has never been in an accident, we understand the speaker's words in an absolutely univocal sense to represent the state of his mind about the car's history. On the other hand, when told by a domestic that someone is not at home, we understand that the language may be a polite way of saying that the party is unwilling to come to the door. In still other contexts, "no" may mean simply "none of your business." Indeed, social convention appears to regard, and to have so regarded throughout history, fictitious statements by government agents in quest of criminals and spies as basically noncommunicative speech. In short, there is considerable variation in the use of language to communicate thought.

Similarly, we can without much reflection conclude from the obligation to live reasonably with others in society that it is morally wrong to kill another human without just cause. But it takes considerably more reflection to decide when or if a particular cause justifies killing another (cf. ST 1–2. 100. 8. ad 3). In the first place, reason can understand on reflection that each human life belongs to its possessor as a matter of right,[11] with the consequence that no human authority, public or private, is morally permitted to take an innocent human life (ST 2–2. 64. 6).[12] But reason can also understand on reflection that humans forfeit their right to continued life when they seriously injure the common good (ST 2–2. 64. 7; 1–2. 100. 8. ad 3). Those who commit serious crimes like murder threaten the very existence of organized society, and public authority may justly inflict death on the evildoers to punish their crimes.[13] Nations which commit aggression against other nations likewise threaten the very existence of the organized societies attacked, and so public authorities of defending nations may justly order their armed forces to inflict death on the armed forces of the aggressor nation to punish the aggression, as we shall argue more fully in a later chapter.

Conclusions about the justifiability of killing humans require of reason considerable reflection on the relation of potential or actual circumstances to first principles. Killing innocents is a different case from those of killing criminals or aggressor enemies of the commonwealth, and killing on private authority is a different case from that of killing on public authority. More important, reason needs to reflect on the relation of human life to its possessor in order to conclude that innocent human life should be inviolable under all circumstances, and reason needs to reflect on the relation of individual human life to the good of the community in order to conclude that public authorities may justly punish with death those who commit serious crimes or armed aggression against the commonwealth.

The conclusion that abortion is contrary to the natural law is a still more remote secondary precept. Because fetuses are incapable of rational activity during part or all of pregnancy, the status of human fetuses as persons can be disputed.[14] But the human life process indisputably begins with fertilization of the ovum, and reason can recognize on reflection that even inchoatively human life, however presently incapable of rational activity, should be respected on that basis alone.

Aquinas would permit no human authority to kill innocents, and he would permit no private person to kill wrongdoers. Both the precept against humans killing innocents and that against private persons killing wrongdoers proscribe direct, that is, intentional, killing of humans. But there are other situations where death results from human action without an intention by the agent to kill another. Such indirect or unintentional killing Aquinas would justify (ST 2-2. 64. 7) on what has come to be called the principle of double effect: one may intend good effects without intending the bad ones.

Moral philosophers and theologians have elaborated four conditions to test whether or not the principle of double effect is satisfied. First, prospective action with both morally good and morally bad effects should itself be morally good or at least morally neutral, that is, capable of producing a morally good effect. Second, the morally good effect should be directly intended, and the morally bad effect foreseen but not desired. Third, the morally bad effect should not be desired as a means toward the morally good effect. Lastly, the morally good effect should be proportional to the morally bad effect, that is, the morally good effect should be of equal or greater weight than the morally bad effect.

Self-defense involving the death of a life-threatening aggressor can satisfy these conditions. Self-protective acts can produce the good effect of physical survival. Human actors can intend to ward off attack rather than the death of the aggressor. Human actors can intend not to inflict death on the aggressor as a means of physical survival.[15] And physical survival is proportional to the death of an aggressor. With respect to the third condition, one might object that a self-defense of one's own life, which is realized by the death of an the aggressor, typically involves that death as the means to physical survival. But with a right intention, one using only the minimum force necessary to resist the aggression need not intend the death of the aggressor in such situations; one sincerely acting on the principle of double effect should be delighted if the aggressor were to recover.

We can also note that application of the principle of double effect to indirect killing does not depend on the personal guilt or innocence of the aggressor for his actions. When humans defend their lives by minimal protective acts of force, they do not seek to punish the other. The principle of double effect, therefore, will justify indirect killing of sleepwalkers or lunatics as much as the indirect killing of willful wrongdoers threatening one's life. (Public authorities, on the other hand, may morally inflict the punishment of death only on those who are subjectively culpable of wrongdoing.) A similar line of double-effect reasoning may be advanced to justify abortions when necessary to save the lives of mothers. As in the case of sleepwalkers or lunatic adult aggressors, the fetus may threaten the life of the mother, of course unwittingly. If so, it would seem that the threat of the fetus to the life of the mother may be legitimately repelled by induced abortions, provided that one does not intend the fetus' death, and abortion represents the minimum force necessary to repel the threat.[16]

Our discussion of natural law precepts against killing has focused thus far on killing acts. We concluded, with Aquinas, that private persons may not directly intend the death of another human, but that they may indirectly permit death if conditions for applying the principle of double effect are satisfied. But failure to act to preserve or save the life of another can result in the other's death and be as morally culpable as a killing act. Parents have a moral duty to provide food for their children, and doctors have a moral duty to heal their patients if they can. Moreover, even in the absence of special relationships like those of parent–child and doctor–patient, there is a moral obligation on the part of all to take reasonable means to preserve or save the life of another. This obligation derives from the love of others toward which humans naturally incline. The converse is also true; there is no moral obligation to do more than is reasonable to preserve or save the life of another. While humans are never morally permitted directly to kill innocents, they are morally permitted to omit acts to preserve or save life when saving acts would be unreasonable. In other words, one may compassionately allow another to die in certain circumstances, and such "letting die" does not violate the precept against killing.

The distinction between killing acts and letting die is of critical importance for correctly understanding moral responsibilities for the lives of those who are terminally ill or born severely deformed. On the one hand, any form of euthanasia is morally wrong because it involves direct killing of innocent human life.[17] On the other hand, compassionately to let another die when it is physically possible to prolong life is morally permissible if and when it would be unreasonable to prolong life. Whether or not acts to prolong life are reasonable will depend on whether or not the quality of incremental life outweighs the risks, pain, or costs of the acts to a patient and his or her family. The burgeoning field of bioethics is concerned with applying that test to particular situations. Here it will suffice to indicate extremes. To continue artificial respiration for a permanently comatose patient like Karen Quinlan would go beyond any moral duty to preserve life, but to provide food for a mentally defective human would be a moral duty.

One purpose of this excursus on lying and killing is to persuade the reader that moral reasoning is a complicated business, and that careful reflection on the relation of contingent events to first moral principles is essential to the correct understanding and formulation of secondary precepts of the natural law. Ordinary citizens, of course, do not typically engage in such reflection and reasoning. (For that matter, professionals may also fail to do so, or at least to do so adequately.) For this reason, public attitudes on issues like "mercy killing" and abortion cannot be considered determinative of the moral questions at stake.

The justice of institutions and economic distributions within organized societies and the justice of relations among organized societies are obviously matters that require of reason considerable reflection on the connection between contingencies and first principles. Considerable reflection is required not only on the ends of organized society but especially on the relation of means to those ends before reason can draw moral conclusions about public policy. Unlike those matters of personal morals which involve elementary life situations, matters of political justice involve very variable circumstances. To formulate public policy in accord with moral principles is, par excellence, the function of professionals.

Not all secondary precepts of the natural law are of the same weight. The precept against killing, for example, concerns a human value considerably more important than those typically at stake in the precepts against stealing and lying. Human life is more important than material goods and truthful statements, unless the latter also involve matters of life and death. Moreover, in matters like stealing and lying, we can distinguish between big thefts and little thefts and between big lies and little lies, whereas no such quantitative distinction is possible in the matter of different human lives.

From our treatment of the precepts against lying and killing, it can be seen that moral reasoning necessarily involves a certain amount of casuistry; cases do differ, and the differences may lead to very different moral judgments about the cases. But the role of casuistry in natural law moral reasoning should not be exaggerated. Sound moral reasoning is primarily concerned with understanding and articulating sound moral principles. There will always be ambiguities when the principles are applied to concrete cases, and so moral reasoning need not be concerned with or anxious about dotting every "i" and crossing every "t." The important thing is to get one's principles straight and to keep first principles clearly in sight. One should not expect of natural law the moral equivalent of the U.S. Code. Indeed, excessive concern to apply rules to cases will make the rules themselves seem heteronomous (extrinsic) to human decision making. Even priest-confessors, whose role requires extensive casuistry to advise penitents on what is sinful in the eyes of the Church, need to be mindful of casuistry's limitations.[18]

Variability and Invariability of the Natural Law

We shall now examine whether or in what sense precepts of the natural law may vary.[19] Are any precepts of the natural law valid under some circumstances or on some occasions but not under other circumstances or on other occasions? And if so, when? We shall first consider whether or when precepts of the natural law may vary in their recognition by persons and peoples. Then we shall consider whether or not precepts of the natural law may vary in their truth-content.

As already indicated, the absolutely first principle of practical reason is self-evident in its accessibility to human reason: all humans assent to the principle on understanding the terms. There is, therefore, no possibility that any human being with the use of reason could fail to recognize that humans should seek what is humanly good and avoid what is humanly bad. Second, with the aid of experience, humans recognize that their specific good includes preservation of one's life, sexual union, family life, search after truth, and life in society with other humans, and they without argument assent to their moral obligation to seek these goods according to rules of reason. There is, therefore, no possibility that any human being with the use of reason and the requisite experiential knowledge can fail to recognize the validity of the primary precepts of the natural law. Third, proximate secondary precepts, like the precept against killing humans without justification, flow so readily from primary precepts that most persons and peoples recognize the validity of proximate secondary precepts; if some do fail to recognize the latter, as Aquinas says, it can only be because passions, bad habits, or customs blind reason (ST 1-2. 94. 4). But remote secondary precepts, like the precept against directly killing innocent humans under any circumstance, require considerable reflection, and so many may fail to recognize remote secondary precepts on that account.

Knowledge of remote secondary precepts of the natural law will vary in different cultures. Reason does not operate simply as an isolated individual effort but within a collective framework of popular culture and historical experience. There are both social and historical dimensions to the operation of reason. As collective perspective on human experience grows, so too does collective perspective on the implications of first principles of the natural law for that experience. A most significant case in point is the growth in the course of Western history of recognition of the value of the human person and of secondary precepts of the natural law reflecting that value. As Jacques Maritain has observed, moral norms like that of monogamy were only recognized rather late in the history of mankind.[20] And recognition of natural law precepts of general political freedom and equality are of even more recent vintage.

Not that collective consciousness of the mandates of the natural law is always on a progressive track. The process of human learning, after all, may be matched by a process of human forgetting. And so, in the area of personal and family values, Western man in the twentieth century has come to disavow some previously recognized precepts of the natural law. Yet, for all the reality of collective "forgetting" of certain norms of the natural law, articulate voices can still be heard to remind their fellows of these norms. In this sense, at least, we Westerners can say that we now have accessible to us more of the natural law than people at any previous moment of history.

Certain situations are intrinsically more complex than others and so require more reflection by professionals in order to reach correct moral conclusions about the situations. This is particularly the case with the political justice of institutions and economic distributions within organized societies and the relations among organized societies. Moreover, public policies may involve highly technical questions on which there may be available only fragmentary data. Thus, even professionals may be and often are unable to agree on the rightness of public policies within the constraints of finite reason and limited information. But the complexity of problems and the limits on information by no means entail that precepts of the natural law about public policies vary in their truth-content. What may vary in complex situations is subjective knowledge or the process of reasoning or both, not the objective truth-content of the conclusions.

Thus far we have examined only the extent to which the recognition of precepts of the natural law may vary from person to person or from people to people. Now we shall consider the more serious question—whether or not those precepts may vary in their truth-content. For example, may a precept of the natural law prohibit an action in certain circumstances but permit the very same action in others? Since it is obvious that no two actions can be exactly the same in every particular, the problem is to determine whether or not different circumstances can so significantly affect the content of natural law precepts that what the precepts themselves command, permit, or prohibit changes. In short, we need to ask whether or not circumstances can change what the natural law prescribes or proscribes.

Unlike the conclusions of speculative reason, which always at least purport to be universally valid, conclusions of practical reason may be so formulated as not to be valid in all cases (ST 1-2. 94. 4). Aquinas cites the example of the secondary precept that property held in custody for an owner should be returned to him on request. That secondary precept is valid in most cases, but it might be unreasonable to return the property in certain circumstances. A secondary precept of the natural law may fail to be universally valid because it is incompletely formulated, as is the case with the formulation that property held in custody for the owner should be returned to him without the proviso "unless it is unreasonable to do so." But if the proviso is explicated, the precept will be universally valid.

Aquinas makes a similar point with respect to the Decalogue prescription against human killing (ST 1-2. 100. 8. ad 3). An objector argues that the precept is not universally valid because it is right for public authorities to inflict death on criminals and enemies of the commonwealth. Aquinas replies that the Decalogue precept against human killing includes the notion of justice, and so human law can justly authorize the killing of criminals and enemies of the commonwealth without violating the Decalogue precept against human killing. Thus, the truth of the secondary precept against human killing, when reformulated to include the notion of justice, will be universally valid.

Some secondary precepts of the natural law thus vary from case to case only because they are incompletely formulated. What of the possibility that secondary precepts of the natural law may vary because of intrinsic changes in the conditions of society? As a concrete example, we shall take up the case of "just" wars to vindicate national rights. Aquinas would permit such wars, but contemporary international opinion would prohibit them.

Although we shall take up this question in greater depth in Chapter 6, let us assume here for the sake of argument that Aquinas correctly articulated a secondary precept of natural law, which would justify wars to vindicate national rights in the thirteenth century, and that the contemporary consensus correctly articulates a secondary precept of the natural law against the justice of such wars in the twentieth century. The truth-content of this secondary precept on "just" wars would thus seem to have changed from one time period to another. But the justice of war necessarily involves at least the notion of proportionality, and the proportionality of means of war to the ends of war is as integral to the justice of thirteenth-century war as to the justice of twentieth-century war. It is the argument of the twentieth-century consensus that the advent of nuclear and other highly destructive instruments of war has transformed previously justifiable wars into unjustifiable wars. Thus, the truth-content of the natural law precept on "just" wars, if formulated to express the condition of proportionality, has not changed.[21] What has changed is the subject matter to be proportioned.

Leading Catholic theologians today take a consequentialist approach to moral norms different from that of Aquinas.[22] This approach derives currency from the Anglo-American Utilitarian tradition wherein consequences determine the moral quality of human acts. But contemporary Catholic consequentialists differ from their secular Utilitarian counterparts in one important particular: Catholic consequentialists recognize consequences as part of an objective, natural order of humanly perfective goods.

Catholic consequentialists maintain that rational apprehension of actions as ontologically good is distinct from rational apprehension of actions as morally good, and that reason apprehends actions only as ontologically good antecedent to the need for action in situations where one ontological good conflicts with that of another. Louis Janssens calls the goodness of objects before the need for action in the face of conflicting goods "ontic," Joseph Fuchs calls it "premoral," and Bruno Schüller and Richard A. McCormick call it "nonmoral." These positions separating rational apprehension of ontological good from rational apprehension of moral good are evidently at variance with the position of Aquinas. According to Aquinas, nature inclines humans towards certain objectives, and reason immediately apprehends these objectives as morally good, that is, as objectives which humans, as rational, should seek, and contrary objectives as morally bad, that is, as objectives which humans, as rational, should avoid. Similarly, reason apprehends other objectives mediately, with reasoning from first principles, as morally good or bad because of their relation to the objectives of natural inclinations.

Contemporary Catholic consequentialists and Aquinas, disagreeing about how reason apprehends actions to be morally good or bad, correspondingly disagree about how reason articulates moral norms to govern action. For the Catholic consequentialist, reason apprehends actions to be morally good or bad only in the context of a concrete choice between conflicting ontological goods, and so no strictly moral norms can exist outside that context; reason, while recognizing and respecting ontological goods, has no need to respond to them except when the need arises to choose between them when they conflict. No doubt Catholic consequentialists wish to avoid universal moral commitment to particular goods before weighing the claims of conflicting goods. But it seems difficult to separate "respect" for naturally perfective goods from moral response to them. If, as Catholic consequentialists admit, reason apprehends the attainment of certain objectives as naturally perfective, then reason should also apprehend actions to achieve the objectives precisely as morally good. Moral good and moral norms thus seem inescapably linked to rational apprehension of objectives and actions as naturally perfective.

Similarly, Catholic consequentialists would not determine the morality of human acts simply by the intrinsic relation of acts to their ends. As these consequentialists would distinguish rational apprehension of ontological good from rational apprehension of moral good, so they would distinguish rational apprehension of the ontological goals of human action from rational apprehension of its moral goals. This view too differs very much from that of Aquinas. In the case of speech, for example, Aquinas argues that lying is morally wrong because communicative speech is intrinsically ordered to representing the speaker's mind to others in society (ST 2-2. 110. 3). When reason appreciates the communicative purpose of human speech and relates that purpose to the natural inclination to live cooperatively in society with others, reason can conclude that truthful communicative speech is morally good, and that falsifying communicative speech is morally bad.

At the heart of contemporary Catholic consequentialism is a methodological refusal to consider any single ontological good as morally absolute, that is, as morally normative without reference to conflicting ontological goods in particular situations. These consequentialists might conclude, for example, that humans should never kill other innocent humans, but their method requires them first to weigh the evil consequences of such killing (e.g., death of the innocent, bad precedent) against the good consequences (e.g., saving other human lives). Such consequentialist reasoning would be very different from Thomistic reasoning on the inviolability of innocent human life. Thomistic thought regards human life as something to be loved simply because it is human and innocent human life as something not to be directly harmed simply because it is innocent.

Catholic consequentialist method would restrict practical reason to a role of calculating the balance between nonmoral good and nonmoral evil. But calculating nonmoral consequences is neither the only nor the most important function of practical reason. Practical reason, without any recourse to argument, apprehends the moral goodness of those objects toward which nature inclines humans, and practical reason with reflection and reasoning understands the moral goodness of other objects because of their relation to the objects of natural inclinations. Thus, the most important function of practical reason is to reflect on the relation of contingencies of experience to primary moral goods and to draw proper conclusions. Right practical reason not only concludes to moral good and moral norms; more importantly, contrary to Catholic consequentialist method, right practical reason also begins with moral good and moral norms.

The Role of Conscience

Applying principles of natural law to concrete action, Aquinas integrated Aristotelian reliance on practical reason with Pauline appeals to conscience.

The word "conscience" is used nowhere in the Gospels. Indeed, it appears only once in the Old Testament, where the Book of Wisdom (17:11), adapting Semitic thought to Greek modes of expression, declares that "wickedness is . . . condemned by its own testimony," and that, "distressed by conscience, it has always exaggerated the difficulties."[23] The writer of Wisdom uses the term "conscience" in the sense of moral consciousness of guilt subsequent to morally bad action.

Although the word "conscience" was not used frequently by Paul, it does appear twenty times in the letters attributed to him and ten times elsewhere in the New Testament. According to Paul, who took the term from contemporary Greek usage, the purpose which the Mosaic law served for the Jews was fulfilled by conscience for the pagans: "what the Law requires is written on their [the pagans'] hearts, while their conscience also bears witness, and their conflicting thoughts accuse or perhaps excuse them on that day when . . . God judges the secrets of men by Christ Jesus" (Rom. 2:15–16). But conscience works in Christians too: "the aim of our charge is love that issues from a pure heart and a good conscience and sincere faith" (1 Tim. 1:5). Indeed, Paul urged Christians to obey the Emperor "not only to avoid God's wrath but also for the sake of conscience" (Rom. 13:5), that is, not only to avoid punishment from imperial authorities but also to avoid the pains of a guilty conscience. Paul cited his own good conscience in preaching the gospel: "I am speaking the truth in Christ, and I am not lying: my conscience bears me witness in the Holy Spirit" (Rom. 9:1). Paul also claimed a good conscience about his apostolic activity up to the present: "I have lived before God in all good conscience up to this day" (Acts 23:1), and "our boast is this, the testimony of our conscience that we have behaved in the world and still more toward you with holiness and godly sincerity" (2 Cor. 1:12).

Of the twenty references to conscience in the letters attributed to Paul, eight appear in Paul's First Letter to the Corinthians in connection with the controversy over eating meat that had been used in pagan sacrifices. There and in Romans, Paul clearly indicated his view that even an erroneous conscience morally binds the individual. Although Paul himself considered it no sin for Christians to eat meat that had previously been offered to idols, he recognized the moral obligation of scrupulous members of the Corinthian and Roman communities not to do so: "I know and am fully persuaded that nothing is unclean in itself, but it is unclean for anyone who thinks it unclean" (Rom. 14:14). Therefore, he cautioned other members of these communities to respect the consciences of their scrupulous brothers and sisters to the extent of themselves not eating meat that had been offered to idols: "if anyone sees you . . . at table in an idol's temple, might not he be encouraged, if his conscience is weak, to eat meat offered to idols? . . . Thus, sinning against your brethren and wounding their conscience when it is weak, you sin against Christ" (1 Cor. 8:10, 12).

Although Paul's uses of the word "conscience" seem to have been linked to negative precepts ("thou shalt nots") and to resulting feelings of guilt when these precepts were breached, his central theme of the supremacy of living faith over external works (cf. Rom. 4) indicates that his moral doctrine was broader than his citations of conscience. There are in Paul, as in the Old Testament, two strands of moral doctrine: one negative, legal, and rule-oriented; the other, positive, prophetic, and oriented toward realization of ideals of justice and love. At any rate, Paul's writings were more influential than any other source in popularizing the concept of conscience in the Western world generally and in contributing to Aquinas' theory of conscience in particular.

In the classical Greek world, the word "conscience" was not common. Democritus of Abdera used the word "conscience" only once in the sense of moral as distinguished from psychological consciousness;[24] Plato and Aristotle never used the term. Socrates did appeal to a divine monitor to justify his actions against the charges that he was corrupting the youth of Athens and undermining belief in the civic gods: "I am subject to a divine or supernatural experience (*theion kai daimonion*), . . . a sort of voice which comes to me," and which, "when it comes, . . . always dissuades me from what I am proposing to do and never urges me on.[25] Whether or not Socrates' citation of a divine monitor is to be equated with a citation of conscience, and this most scholars deny, Socrates at least did not use the term "conscience" to describe his divine monitor. It was rather from the Stoics that the term "conscience" came into philosophical parlance. Cicero, for example, wrote that "the consciousness (*conscientia*) of living well and the record of deeds done well is most pleasing."[26] And Seneca expressed the idea of conscience as "a scared spirit in man," which is "an observer and guardian of good and evil in us."[27]

Yet, it was the ethics of Aristotle, not that of the Stoics, which principally influenced Aquinas' philosophical analysis of conscience, and Aquinas relied on Aristotle's concept of practical reason to interpret Pauline appeals to conscience. While Aristotle did not use the term "conscience," his practical reason played a similar role with respect to concrete action. "Practical wisdom," Aristotle says, "issues commands: its end is to tell us what we ought to do and what we ought not to do" (NE 6. 10. 1143[a]7–9).[28] Aquinas transforms this concept of reason ordering action to the moral good of human well-being into a concept of reason applying moral law to action.

In the first part of the *Summa Theologica*, Aquinas argues that conscience is an act, a judgment of practical reason with respect to a particular action. The term "conscience" etymologically means the reference of knowledge to something else (*cum* [*alio*] *scientia*), and "the application of knowledge to something is done by some act" (ST 1. 79. 13).[29] According to Aquinas, this application of knowledge to action can occur in three ways: (1) if we recognize that we have or have not done something, then conscience is said to witness; (2) if we judge morally that something should or should not be done, then conscience is said to urge or bind; (3) if we acknowledge that something has been done morally well or ill, and then conscience is said to approve or accuse. Because habits dispose individuals to acts and so facilitate the acts, Aquinas admitted that the term "conscience" can be applied by extension to the habit of understanding first principles of action, *synderesis*, as well as to the act of judgment itself.

The application of knowledge to action in the first way described by Aquinas identifies conscience with psychological consciousness, and this accurately reflects the first meaning of the Latin term "*conscientia*" and its Greek equivalent, "*syneidesis*." The second way applies knowledge to the moral quality of prospective action and morally binds the individual. The third way applies knowledge to the moral quality of past action and results in feelings of guilt or innocence. The third way obviously depends on the second, and the second on the first. It was with the second way, however, concomitant knowledge of the moral quality of present action, that Aquinas was primarily concerned in his treatment of conscience.

When Aquinas deals with the moral quality of human acts in the second part of the *Summa*, he asks first whether or not the human will is morally bad when it acts at variance with an objectively erroneous judgment of conscience, and he answers that "absolutely speaking, every will at variance with reason, whether right or erring [reason], is always evil" (ST 1-2. 19. 5).

In the next article of the same section, Aquinas asks whether or not the human will is morally good whenever it abides by erring reason. This is a different question from the one asked in the preceding article. In article 5, Aquinas asked only whether or not the human will acts wrongly if it acts *against* a judgment of erroneous conscience. But in article 6, Aquinas asks whether or not the human will acts rightly whenever it acts *in accord with* a judgment of erroneous conscience. The former article asked whether or not a judgment of erroneous conscience that commands or prohibits an action morally binds the individual in error; the latter article asks whether or not a judgment of erroneous conscience that permits an objectively prohibited action or omission of an objectively obligatory action morally excuses the individual in error.

The answer to the question whether or not an erroneous conscience excuses, Aquinas argues, depends on whether or not the agent is responsible for the fact that his or her conscience is in error. If the agent is not responsible for the ignorance leading to an erroneous judgment of conscience, ignorance will excuse action in accord with the judgment. But if the agent is in any way responsible for the state of ignorance, ignorance does not excuse action according to the erroneous judgment of conscience. Applying this distinction, Aquinas concludes that errors of fact may excuse actions in accord with erroneous judgments of conscience but not "ignorance of the divine law, which he [the individual in error] is bound to know" (ST 1-2. 19. 6).

Aquinas presumably does not refer here to ignorance of primary precepts of the natural law because he holds that all humans with the use of reason do in fact assent to those precepts. Aquinas also presumably does not refer to ignorance of remote secondary precepts of the natural law because he holds that the latter are recognized only after considerable reflection. And Aquinas presumably does not refer to ignorance of divine positive law because he holds the faith necessary to recognize divine positive law to be a free gift from God. If these presumptions are correct, then Aquinas is reductively saying that any ignorance of proximate secondary precepts of the natural law, those arrived at with little reflection, will not excuse action according to an erroneous judgment of conscience. Because ignorance of proximate secondary precepts of the natural law is easily avoidable, any action contrary to the precepts is indirectly voluntary, and so humans are responsible for any morally prohibited act they perform and any morally obligatory act they omit in a state of such ignorance.

From the foregoing sketch of Aquinas' doctrine on conscience, several points stand out. First, humans are inclined by nature to seek moral good and to avoid moral evil. Second, nature endows humans with first principles of practical reason, and humans have a natural habit of understanding them. Third, conscience in the strict sense is an act, a judgment of practical reason that a particular action should or should not be done. Fourth, judgments of conscience commanding or prohibiting actions should be followed even if erroneous. Fifth, erroneous judgments of conscience about the permissibility of actions commanded or prohibited by proximate secondary precepts of the natural law will not be excused. Sixth, Aquinas conceived the operation of conscience in a theocentric context of human responsibility to divine and natural law. Lastly, contrary to much contemporary usage, conscience applies moral norms, not moral ideals, to human behavior.

Notes

1. I follow here the path so well laid out by David Brown, *Choices: Ethics and the Christian* (Oxford: Blackwell, 1983), pp. 1–24, to indicate the chief alternatives to natural law ethical theory.

2. Joseph F. Fletcher, *Situation Ethics* (Philadelphia: Westminster Press, 1966).

3. See, e.g., Gustavo Gutiérrez, *A Theology of Liberation*, trans. and eds. Caridad Inda and John Eagleston (Maryknoll, N.Y.: Orbis, 1973). I make no claim that Gutiérrez or any other liberation theologian holds the views ascribed to them here. For illustrative purposes, I simply assume their critics' perspective.

4. John C. Bennett, "Cultural Pluralism: The Religious Dimension," *Social Order*, February 1961, p. 57.

5. The use of the term "teleological" here to describe Aristotle's goal-oriented philosophy should be carefully distinguished from the usage in contemporary ethics. Aristotle's philosophy is called teleological because things and activities are ordered intrinsically to ends set by nature. In contemporary ethical usage, a teleologist is one who holds the view that consequences, i.e., the net value of

nonmoral good over nonmoral evil, constitute the ultimate moral criterion of human action. See William K. Frankena, *Ethics*, 2nd ed. (Englewood Cliffs: Prentice-Hall, 1973), pp. 14–15. A deontologist is usually defined by contrast as one who holds the view that moral norms do not always depend on the balance of consequences in particular cases. Frankena, p. 17. For reasons given in the text, it is not entirely clear how Aristotle would fit into contemporary teleologist-deontologist usage, although Aquinas would qualify as a deontologist insofar as the latter claims that secondary precepts of the natural law can be so formulated as to be universally valid without regard to particular circumstances.

6. The reference here to "other" activities is not intended to imply that activities of reason themselves are not subject to the norm of right reason. As with other activities, activities of reason will only be reasonable at certain times, and in certain places, and under certain circumstances.

7. In the play itself, repeated references to the command of God, the holy, sacrilege, etc., indicate that Sophocles' Antigone herself appealed to an explicit divine command of piety rather to any purely natural law, however pregnant the theme for future development of natural law theories.

8. John Finnis, *Natural Law and Natural Rights* (Oxford: Oxford University Press, 1980), pp. 94–95, rejects Aquinas' three-fold order of primary precepts as "irrelevant schematization" (p. 95). But those principles are both relevant and necessary to substantiate any attempt, like Finnis' own, to defend basic values that are not to be sacrificed. One can only conclude, for example, that humans should never directly take innocent human life if humans have a natural moral obligation to act sociably toward others, and if the direct killing of any innocent would be to act unsociably toward that other. Finnis undermines his project to establish seven basic values that humans should respect in every act by failing to ground the values on the threefold inclinations of human nature itself. Without such a foundation, Finnis' list of basic values will appear to many to be idiosyncratic.

9. Aquinas does, however, give a rule of thumb to help identify those secondary precepts which require little reflection to recognize and those secondary precepts which require greater reflection: the more weighty and repugnant to reason the objects of human actions are, the more easily reason can apprehend them to be morally bad. ST 1-2. 100. 6.

10. Aquinas considers the Decalogue precept of Sabbath observance here a precept of natural law only insofar as it commands that men set aside some time for divine worship, not insofar as it commands specifics of Sabbath observance, ST 1-2. 100. 3. ad 2.

11. Human personality, i.e., human rationality and freedom, is the foundation of human rights. Animals, which have no rationality or freedom, have no rights to life or anything else. But from this it by no means follows that humans may do whatever they wish to animals. Cruelty to animals, for example, is wrong, not because animals have rights, but because humans have reason. To inflict purposeless pain on animals would be to act against *human* reason.

12. Aquinas is careful to make clear that human life can be said to be innocent in relation to other human beings, not in relation to God. As a consequence of original sin, God can inflict death on any human being at any time, and so Aquinas would permit human agents directly to kill "innocents" at the command of God. ST 1-2. 94. 5. ad 2. For a definitive study of the texts of Aquinas on this matter, see Patrick Lee, "Permanence of the Ten Commandments: Saint Thomas and his Modern Commentators," *Theological Studies* 42 (September 1981): 422–43.

13. When Aquinas says that public authorities may justly execute criminals as punishment for their serious crimes against the commonwealth, he is not

necessarily commending the policy or institution of capital punishment. Capital punishment may be wrong for reasons other than the right of convicted criminals to life. The institutionalization of capital punishment may result in the death of those who are in fact innocent, inflict unnecessary pain on the guilty in the course of execution, and/or brutalize the public.

14. Aquinas himself, following Aristotle, held that human fetuses became persons with the infusion of human souls before the beginning of the second trimester of pregnancy (males earlier than females!). *De veritate catholicae fidei* 2. 89. Although Aquinas does not explicitly treat the morality of abortion in the *Summa*, it is clear that he would regard abortions after infusion of the human soul, i.e., during the second and third trimesters of pregnancy, as subject to the precept against direct killing of innocents, and abortions during the first trimester of pregnancy as subject to the precept against contraception.

15. Not all Thomists follow Aquinas on this point. Those disagreeing argue that persons defending themselves against a life-threatening aggressor in fact do, and in justice may, directly intend the aggressor's death in such circumstances. See, for example, Juan de Lugo *De justitia et jure* 10. 6. 14. That view seems to accord with common sense judgment. One might also argue that public authorities sanction the death of life-threatening aggressors, and so that individuals killing aggressors in self-defense are not purely private agents. In any case, both Aquinas and de Lugo agree that humans may never directly intend the death of innocents, and that the indirectly resulting death of innocents needs to satisfy the principle of double effect.

16. Several nineteenth-century Roman Catholic moral theologians took this position. See Jean P. Gury, S. J., and Antonio Ballerini, S. J., *Compendium Theologiae Moralis*, 2d ed., 2 vols. (Rome: Typis Civitatis Catholicae, 1869), 1:326–31, and August Lehmkuhl, S. J., *Theologia Moralis*, 2 vols. (Freiburg: Herder, 1883–84), 1:499–500. Whatever the merits of the position, it was certainly not refuted by the lapidary response of the Holy Office in 1895. Heinrich Denziger and Adolf Schönmetzer, eds., *Enchiridion Symbolorum* (Freiburg: Herder, 1965), n. 3298, p. 643. Nor was the position refuted by the conclusionary statement in *Casti Connubii*, Denziger-Schönmetzer, n. 3720, p. 725.

17. I use the term "euthanasia" exclusively to refer to killing acts. It is currently commonplace, however, to use the term more broadly to include both killing acts (positive euthanasia) and "letting die" (negative euthanasia).

18. Post-Tridentine Catholic moral theologians attempted to codify moral rules in a comprehensive way. The principal reason for this emphasis on concrete cases was practical rather than theoretical; the teaching of moral theology was tailored to the seminary training of priest-confessors. Such casuistry was not without its practical uses, but it also was not without potential theoretical costs. Overemphasizing application of rules to cases could obscure the necessity of reasoning from the apprehension of primary moral goods to the formulation of secondary natural law precepts.

19. For an excellent treatment of the question of natural law variability, see Ross A. Armstrong, *Primary and Secondary Precepts in Thomistic Natural Law Teaching* (The Hague: Nijhoff, 1966), pp. 143–79. Armstrong examines relevant texts of Aquinas in detail.

20. Jacques Maritain, *On the Philosophy of History*, ed. Joseph W. Evans (New York: Scribner, 1957), p. 105. Maritain there (pp. 104–11) elaborates on the historical development of moral consciousness.

21. Armstrong (pp. 178–79) summarily suggests otherwise. But he does not explore the possibility that a precept permitting "just" wars would include an implicit condition of proportionality. Armstrong himself (pp. 167–68) recognizes

an implicit condition of reasonability in such other precepts of the natural law as that prohibiting the killing of humans and that commanding the return of property held in custody to the owner.

22. See Joseph Fuchs, S. J., *Responsabilità personale e norma morale* (Bologna: Dehoniane, 1978); Louis Janssens, "Norms and Prenotes in a Love Ethic," *Louvain Studies* 6 (Spring 1977): 207–38; Richard A. McCormick, S. J., *Ambiguity and Moral Choice* (Milwaukee: Marquette University Press, 1973); Bruno Schüller, *Die Begründung sittlicher Urteile* (Düsseldorf: Patmos, 1973). McCormick is also the author of annual "Notes on Moral Theology" in *Theological Studies*.

23. This and other scriptural quotations are from the Revised Standard Version.

24. Hermann Diels, *Die Fragmente der Vorsokratiker,* ed. Walther Kranz, 10th ed., 3 vols. (Berlin: Weidmann, 1960), 2:206–07.

25. Plato *Apology* 31C, D; the translation is that of Hugh Tredennick from Plato, *The Last Days of Socrates* (Harmondsworth, England: Penguin, 1959), pp. 63–64.

26. Cicero *De senectute* 3. 9.

27. Seneca *Epistulae* 41. 1.

28. This and other translations of NE are from Aristotle, *Nicomachean Ethics,* trans. Martin Ostwald (Indianapolis: Bobbs-Merrill, 1962).

29. This and other translations of ST are from Thomas Aquinas, *Summa Theologica,* trans. English Dominican Fathers, 2d ed. rev., 3 vols. (New York: Benzinger, 1947).

2

The Human Person and Organized Society

In Chapter 1, we traced Aquinas' moral theory; in this chapter, we shall consider his political theory and compare it with other theories. Aquinas' moral and political theories are integrally related. As with Aristotle, so with Aquinas, no moral theory is complete without a political theory, and no political theory adequate without a moral base. Following Aristotle, Aquinas held that the rational, hierarchical organization of society was natural to humans, with limited government at the apex. This view differs from that of Hobbes, for whom organized society was solely a matter of convention, and unlimited government the means to control the consequences of naturally antagonistic behavior by individuals. The Thomistic view also differs from that of Locke, for whom organized society was a matter of convention, and limited government the means to control the consequences of conflict among individuals. And the Thomistic view differs from that of Marx, for whom government would become superfluous in the reconciliation of self-interests in a properly reconstructed society. We turn now to each of these theories.

Aquinas

Aristotle says that nature itself constitutes humans as "political animals" (*Politics* 1. 2). Aquinas prefers to say that nature constitutes humans as social as well as political animals (ST 1-2. 96. 4;

cf. ST 1-2. 72. 4). There is less difference here than meets the eye. The adjective "political" derives from the *polis*, the Greek city-state, and life in the *polis* comprised the whole network of social relationships. For medievals and moderns, on the other hand, the adjective "political" has a narrower meaning that relates to government or government-surrogates. And so Aquinas correctly translates the Greek adjective "political" as "social and political" for his (and our) contemporaries.

We have already seen in Chapter 1 Aquinas' views that humans are inclined by nature to associate with other humans, and that human reason can recognize a duty to do so on reasonable terms. The first level of human society is the family. Because humans naturally incline to mate and procreate, they associate in nuclear families, and nuclear family units associate to form extended families by reason of their blood and marriage relationships. Out of these extended families come clans and tribes.

In addition to social relationships based on blood and marriage, there are purely voluntary human associations. Economic activity brings humans into association, whether businessmen with other businessmen, employees with other employees, or employers with employees. Like interests will also lead humans to form baseball leagues, bridge clubs, and discussion groups.

In an early work (DR 1. 1. 5-7),[1] Aquinas argues that the social cooperation of family units, with a division of labor, is necessary for human survival and material well-being.[2] In later works (CE 1. 1. 4; cf. ST 1-2. 72. 4), Aquinas follows Aristotle more closely and argues that social cooperation is necessary to achieve the specifically human goals of intellectual and moral well-being. Over and above mere economic subsistence and convenience, humans, because of their rational nature, desire to grow in knowledge and virtue, and only by and through human associations can they do so.

But neither families nor small groups of families can adequately provide for either their material or their nonmaterial well-being. They cannot adequately provide for their physical security. They cannot adequately provide for their economic sufficiency and development. They cannot provide a broad base for the development of human friendships. Only a larger society, a politically organized society, can do this, and Aquinas accordingly designates that politically organized society "perfect" in contrast to other social units like the family (ST 1-2. 90. 2).

The body politic arises when families and groups of families see themselves as constituting one people with a common goal of human development. Ethnicity provided the foundation for the body politic in the ancient world. But ethnicity is not a necessary condition for bodies politic in the modern world, although it is still often associated with them. There are many modern societies composed of different ethnic groups, of which the U.S. is perhaps the most conspicuous example. Nor is ethnicity a sufficient condition for the formation of a body politic. The body politic only comes into existence when otherwise disparate families and clans recognize common goals, not merely common origins. In other words, the body politic represents a rational organization of social relationships, not merely an affective orientation toward them.

The rational organization of social relationships implies notions of state and government. The body politic consists of individual family units recognizing a common goal of human development. But the means of achieving the common goal of human development need to be specified and determined. Societies will need to specify acceptable and unacceptable modes of behavior and to punish deviants; every rationally organized society will need a criminal code. Other specifications of means to human development will be necessary without reference to the virtue or vice of individual members of society; a rationally organized society will need to regulate aspects of civil life even for the virtuous members of society. And modern societies will need more regulation than primitive or traditional societies. The institutional agency responsible for specifying the common good of the body politic is the state. The state—and only the state—has the function of specifying that good. The institutional machinery of state may be simple, as in the case of a tribal council, or complex, as in the case of separate legislative, executive, and judicial branches of government. The term "government" refers to the offices of state or to the office-holders.

The purpose of the state is to promote the common goal of developing human personality both materially and spiritually. The overall goal of human development is itself specified by the structure of the human person. Put another way, the state exists to organize human society rationally, and rationality includes the rationality of ends as well as the rationality of means to ends. The state in Thomistic theory, though hierarchically supreme in organized society, is not an absolute arbiter of morality.

Aquinas links human laws essentially to natural law: human law is either a conclusion of natural law (e.g., theft is a crime) or a further determination of natural law principles (e.g., the punishment for theft will be ten years imprisonment) ST 1-2. 95. 2). This linkage of human law to natural law is absolutely essential if human law is to qualify as law at all. "Every human law has just so much of the nature of law as it is derived from the law of nature," and "if in any point it deflects from the law of nature, it is no longer a law but a perversion of law." In short, an unjust law is no law in the true sense.

A second point about the Thomistic theory of the state should also be stressed.[3] The state has hierarchical supremacy over other units of organized society because of its specific function to care for the common good. Nevertheless, the state remains only an instrument or agency of the body politic. The state is only a part, albeit the supreme part, of the whole body politic, the community of human persons. This instrumentalist theory of the state is diametrically opposed to the absolutist theory of the state. According to the latter theory, the state is a moral person, a subject of rights, which totally swallows up the community of human persons, the body politic.

The notion of personality, however, can be properly, if only analogously, applied to the people as a social whole because the people as a social whole constitute a unity of real, individual persons, and because this unity derives from the minds and wills of real, individual persons. Accordingly, the notion of moral or collective personality can be applied in a proper sense to the body politic, which is the people as an organized social whole. The people as a whole have a right to determine their form of government, and there are relationships of justice between the body politic and its members. But the same notion of moral personality cannot be applied to the state except in a metaphorical sense. The state is not the people as a whole but an instrument of the people organized into a social whole. Rights ascribed to the state are really rights of the people in their collectivity.

Bureaucracies generally tend to grow beyond their proper limits simply because power tends to seek more power. General staffs tend to mistake their own good for that of the whole army, and church leaders tend to mistake their good for that of the whole church. State bureaucrats are no different; they tend to mistake their good, their survival and growth, for the common good of the body politic, namely, human development of the persons comprising the body politic. But the absolutist theory of the state involves much more than this tendency of bureaucracies to confuse ends and means, and it can only be understood in the light of modern history. The Baroque era absolutized the state and identified it with the person of the king. The French Revolution continued to absolutize the state but substituted the acephalous people for the king. Twentieth-century totalitarianisms—Nazism, Fascism, and Communism—revealed the absolutist theory's true face.

We should recognize not only that identification of the state with the people as a whole is just as compatible with totalitarianism as with nominal democracy, but also and especially that real democracy is possible only under the instrumentalist theory of the state. We should keep in mind that the state is an instrument, albeit the supreme instrument, of the body politic to order the activities of citizens to the common good of human *self*-development. Only with that consciousness will the state act rationally to supervise the organization of society, as that supervision becomes even more necessary to human development in the complexities of modern society, without threatening the subsidiary societal functions of the human person. Only with that consciousness will it be possible to strike a balance between the state's recourse to coercive power and the freedom of individual citizens. Only with that consciousness will the state and the rule of law be respected; the state will then earn respect for its exercise of justice, not for the trappings of power and prestige.

The body politic, the people organized into a social whole, has subordinate social units as parts. First and foremost are the family units. These social units, though subordinated to the whole and integrated into the whole, retain essential rights and freedom within a rightly ordered body politic. This is so because the purpose of the body politic is human development, and familial societies are integral to human development. Families are natural to humans, that is, nature inclines humans to mate, procreate, and educate their children. Similarly, other human societies that proceed from the free initiative of individuals should enjoy as much autonomy as possible precisely because humans develop themselves by their free associations with one another. Thus, in the Thomistic theory, pluralism is inherent in every rightly ordered political society. The body politic and its chief agency, should not swallow up the parts, families and voluntary associations, but should function to order the parts into a whole, while preserving the maximum autonomy of the parts. This relationship of the parts to the whole and the whole to the parts has an all-important consequence for the state as the supreme agency of the body politic, the principle of subsidiarity: the state should do for its citizens, families, and voluntary associations only what these individuals and groups cannot or will not do for themselves. In other words, the freedom of citizens and their associations to act as they wish should be restricted only as much as the common good requires.

Societies organize themselves politically under different regimes or forms of government. In the treatise *On Kingship* (DR 1. 1. 10–11), Aquinas divides the forms of government into the traditional Aristotelian categories of monarchy, aristocracy, and polity.[4] The basis of the division is simply a matter of numbers. Monarchy is rule by one, aristocracy rule by a few, and polity rule with many participating. These just, that is, rightly ordered, forms of government are directed toward the common good; they are perverted into unjust forms of government when the one governs for his private good (tyranny), the few for their private good (oligarchy), or the many for their private good (democracy).

In later writings (CE 8. 10; ST 1-2. 95. 4 and 105. 1), Aquinas changed focus, as did Aristotle in the *Politics* (3. 8-9), from the number of persons participating in the ruling power to the characteristic virtues according to which political power is awarded in different forms of government. The just forms of government award political power on the basis of virtue: monarchy and aristocracy are given to one or a few on the basis of their virtue of wisdom or practical reason, and, likewise, polity is awarded to the many on the basis of their virtues of courage and moderation. The unjust forms of government award political power on some basis other than virtue: tyranny to one on the basis of physical force alone, oligarchy to a few on the basis of wealth alone, or democracy to the many on the basis of freedom and equality alone.

In the treatise *On Kingship* (DR 1. 2. 17), Aquinas affirms his preference for monarchy as the best form of government because it best promotes unity. But to prevent tyranny, monarchy should be so structured that the opportunity to tyrannize is removed, and the monarch's power so "tempered" that he cannot easily lapse into tyranny (DR 1. 6. 42). In the *Summa* (ST 1-2. 105. 1). This "tempering" of royal power is spelled out in greater detail:

> Accordingly, the best form of government is to be found in a city or kingdom in which one man is placed at the head to rule over all because of the preeminence of his virtue, and under him a certain number of men have governing power also on the strength of their virtue; and yet a government of this kind is shared by all, both because all are eligible to govern, and because the rulers are chosen by all. For this is the best form

of polity, being partly kingdom, since there is one at the head of all, partly aristocracy, insofar as a number of persons are set in authority, partly democracy, that is, government by the people, insofar as the rulers can be chosen from the people, and the people have a right to choose their rulers.[5]

Where Aristotle preferred a constitutional regime that blended oligarchy and democracy, Aquinas prefers a constitutional monarchy that blends monarchy, aristocracy, and democracy. Each preference reflects, at least in part, the author's cultural milieu.

Aquinas derives government from the consent of the people. The right to make law, says Aquinas (ST 1–2. 90. 3), belongs to the whole people or to its representative, a public personage who has the care of the whole people. And Aquinas subordinates statutory law to constitutional custom.[6] For a free people able to make its own laws, that is, an unconquered people, the consent of the people as expressed by custom counts far more than does the authority of the sovereign, "who has not the power to frame laws except as representing the people" (ST 1–2. 97. 3. ad 3). Thus, the sovereign represents the will of the people as embodied in constitutional customs and traditions; the sovereign derives his power from the constitution, not the constitution its power from the sovereign.

Aquinas' political theory is basically Aristotelian. Like Aristotle, Aquinas defines the aim of rightly ordered regimes in terms of the common good. Because humans cannot adequately achieve material, intellectual, or moral well-being apart from their fellows, they form a body politic to promote this common good. Also, like Aristotle, Aquinas envisions the ruler not only as a maker of law but also as a subject of law; the power to rule a free people is constitutionally limited.

Aquinas, however, differs from Aristotle in several important respects. In the first place, unlike Aristotle, Aquinas relates human law to a natural law implicit in God's creative plan, and this natural law places moral limits on human law. For Aristotle, rightly ordered political power should be exercised under a rule of law, that is, in a constitutional framework. For Aquinas, rightly ordered political power should be exercised within the framework of both a divinely ordained natural law and a humanly established constitutional law, that is, both "sub Deo et lege," as the English jurist Henry Bracton put it.[7]

Secondly, Aquinas' archetypical regime differs from that of Aristotle. Where Aristotle focused his reflections on politics around the democracy of ancient Athens, Aquinas focuses his around the hereditary monarchy of medieval Europe.[8] Aquinas could not imagine a Europe without hereditary monarchies as archetypical, and in this respect, his political theory is culturally conditioned. But it is important to add that, although Aquinas accepted the hierarchical class system of medieval society, he was no defender of class privilege for its own sake. Aquinas insists that every just social and political system should be directed toward the good of the community. Justice is for Aquinas both the moral end of political organization and the pragmatic means of political stability.

Beyond justice, both Aristotle and Aquinas understood well the necessity of moral virtue generally for the right ordering of the political community. Marx, in his own way, understood this too. Aristotle and Aquinas recognized that participatory democracies make more demands on the virtue of citizens than any other regime. The democratic principle of equal participation, without morally responsible citizens, risks anarchy. For most of modern Western history, statesmen have concentrated on building and maintaining political structures that would make the world safe for democracy. The question for the West today may be whether or not citizens can be virtuous enough to make democracy safe for the world. Whatever the cultural limitations of Aquinas' political theory, he at least recognized the essential relationship between virtue, both as end and as means, to regime.

A word should be added about the relation of moral principles to statecraft or public policy making. We have already noted in Chapter 1 the complexity of applying primary principles of natural law to human contingencies on matters of political justice within and among organized societies. There we were concerned with showing that public policies have a moral dimension. Here we note the converse. Public policy involves estimates about contingent physical events and human behavior. Reasoning from moral principles, though necessary, is insufficient for sound public policy, and the statesman's application of moral principles to concrete situations depends on empirical contingencies. Statecraft in the Thomistic framework requires both an understanding of normative principles and an estimate of empirical probabilities. The practical judgments that statesmen make about public policy are not so much true or false as they are good or bad, and only experience can show which they are.

Aristotle thought that Greek city-states were relatively self-sufficient bodies politic, and Aquinas thought medieval kingdoms to

be the same—territorially limited political units capable of achieving the material, intellectual, and moral purposes of human society. Various political societies would have commercial and cultural intercourse, and relations among them would be regulated by
natural moral norms and the customary law of nations.

Whether or not such territorially limited self-sufficiency was
possible even in premodern times, it can be strongly argued that in
the modern world national bodies politic are incapable of the self-
sufficiency envisioned by Aristotle and Aquinas.[9] There are two
basic reasons for this. The first is that national economies today
are closely interrelated and interdependent; technology has so rev-
olutionized commerce that no nation can adequately develop its
own economy without trade with other nations. The second and
more critical reason is that no nation today can, even with a
network of alliances, guarantee lasting peace. Indeed, nations
never could, but modern weapons of war, especially nuclear weapons, threaten incalculable or even total destruction should war
occur. Only a worldwide political society and government could
guarantee world peace.

Nations in the twentieth century have widely recognized the
inability of their bodies politic severally to secure lasting peace.
Most nations joined together after World War I to form the League
of Nations and after World War II to form the United Nations. But
the League of Nations was, and the United Nations is, a confedera-
tion of nations, not a world government. Decisions of the United
Nations lack efficacy not only because the U.N. has no armed
forces of its own to coerce member states to obey, but also and more
important because member states do not recognize the moral authority of the U.N. to command unless the command coincides
with member states' own wishes or sense of responsibility.

Recognition of the moral authority of the decisions of an
international organization involves recognition of the organization as a government or organ of government. Before that recognition can be achieved, peoples of the world need first to recognize
themselves as forming one worldwide people. There must be a
common mind and will to live and work together to build a just
world society, not a common mind and will to live apart in a state
of armed neutrality. Living and working together for justice involves sacrifice of national self-interests to the common good of
peoples of all nations. Peoples of the world are currently unwilling to do this, and so no world body politic can exist for a world
government to govern.

Thus, the cards are stacked against the emergence of a world society and a world state in the short or even the long run. The main reason is simply stated. Marxist-Leninist regimes would be unwilling to enter a worldwide political society except on Marxist-Leninist terms, and democratic regimes would be unwilling to enter such a society except on democratic terms. Even if Marxist-Leninist societies did not exist, there is no evidence that democratic national societies are prepared to transcend short-term national interests to live and work together for the long-term common good of the human race.

The rest of the world can do little or nothing to alter Marxist-Leninist ideology, and so a universal world society and world state are impossible to achieve in the foreseeable future. Democratic national societies might be tempted to aspire to a half-universal world society and world state if they could transcend their present identification with national self-interests. Such a half-universal world society and world state, however, would be destabilizing in the current world situation. Marxist-Leninist regimes would undoubtedly and understandably perceive a democratic half-world state as a conspiracy of the rest of the world against themselves. This would vastly increase East-West tensions, making a world war more rather than less likely.

What, then, can democratic societies do to further the formation of a world society and world state in the distant future? First, democratic societies can educate themselves to think beyond their national self-interests. This will involve a conscious rejection of nationalism and all its works. Second, democratic societies can work together to build a cultural world community including Marxist-Leninist societies to the extent that is possible. Third, democratic societies can cooperate in regional economic associations like the Common Market. Finally, democratic societies can work to strengthen existing international organizations. Democratic societies cannot force Marxist-Leninist societies to accept decisions by international organizations like the International Court of Justice, but they can accept the Court's jurisdiction over disputes among themselves.

Hobbes

Aristotle and Aquinas intimately linked what moderns call private morals and public duty; there could be no morally good individual who was not also a politically dutiful citizen and no politically dutiful citizen in the full sense who was not also a morally good person. Consequently, both Aristotle and Aquinas were concerned with the right ordering of the individual and the commonwealth, that is, the structuring of the behavior of both individual and commonwealth according to an objective order of moral good.

With the Enlightenment this view of morals and politics changed radically. At least two major historical forces paved the way for this change. First, the Reformation shattered the religious unity of Western Europe, and this fragmentation of religious unity, in the context of the then-prevalent concept of sacral society, shattered the political unity of Western Europe as well. Catholics linked political allegiance to a regime favoring Catholics, Protestants linked political allegiance to a regime favoring Protestants, and neither sect acknowledged political allegiance to a regime favoring the other. But if the Reformation undercut the religious basis of political organization for Catholic and Protestant dissenters, the empiricism of modern science seemed to undercut the rational basis of political obligation for any citizen. Modern scientific method uses reason to formulate and test empirically verifiable hypotheses. If this were the exclusive method whereby human reason could arrive at truth about humans and their world, then human reason obviously would be unable to apprehend any objects, immediately or mediately, as naturally good, that is, as objects that should be sought because of their relation to a natural order. And, if reason cannot apprehend what is naturally good for humans, then reason is equipped only to calculate the means to achieve what individuals desire. Thus, the Reformation, and especially the new philosophies based on modern science, threatened the moral basis of politics in Western society.

In this section, we shall trace the attempts of Hobbes to restore to politics a moral order that would be consistent with the philosophies based on modern science. This would not be a moral order accessible to human reason before any operation of the will but a moral order resulting precisely from the operation of human will, the social contract.

Hobbes was consciously concerned with elaborating a political theory that would morally oblige citizen-subjects to obey their sovereign and disallow any moral obligation to disobey the sovereign on the basis of private judgment or religious belief. In the state of nature, that is, where no commonwealth exists, humans are driven by competition, distrust of others, and the desire of glory to war with one another. This is a sorry state of affairs. But "reason suggests convenient articles of peace, upon which men may be drawn to agreement," and "these articles are they which otherwise are called the laws of nature" (LN 13).[10] A law of nature is "a precept or general rule found out by reason by which a man is forbidden to do that which is destructive of his life or takes away the means of preserving the same and [is forbidden] to omit that by which he thinks it may be best preserved" (LN 14). Accordingly, "natural right" is "the liberty each man has to use his own power as he will himself, for the preservation of his own nature, that is to say, of his own life, and consequently, of doing anything which in his own judgment and reason he shall conceive to be the aptest means thereunto" (LN 14).

Before considering how the "laws of nature" are articles of peace, we need to note how the state of nature, the laws of nature, and natural rights, as understood by Hobbes, differ from the positions of Aristotle and Aquinas. First, the state of nature postulated by Hobbes separates humans as individuals from humans as members of society. For Hobbes, humans are structurally asocial and apolitical, even antisocial and antipolitical, while for Aristotle and Aquinas, humans are structurally social and political. Hobbesian man is by nature purely aggressive and competitive. For Hobbes, humans are inclined to conflict with others, while for Aristotle and Aquinas, humans are inclined by reason to cooperate with each other.

Second, the Hobbesian laws of nature are rules of reason which forbid an individual to do anything destructive of his life and command him to do whatever he thinks most efficient to preserve his life. For Hobbes, humans are inclined by nature to preserve their lives and to follow reason in their choice of efficient means thereto; for Aquinas, humans indeed naturally incline to preserve their lives, but humans also naturally incline to do so only by those means rationally appropriate to human nature. Hobbes has reduced the natural inclinations of humans to an inclination to survive physically and reduced reason to the role of calculating efficient means to secure survival.

Third, Hobbesian natural rights are rights to do whatever reason calculates to be most efficient to save one's life. For Hobbes, natural rights are unrelated to an objective order of nature, save the natural inclination to preserve one's life, while for Aquinas, natural rights derive from a comprehensive, objective order of nature not only for humans as living substances and animals but also for humans specifically as rational. Hobbes' laws of nature are only dictates of a calculating, self-centered reason. Every human instinctively pursues the goal of securing his or her life. But humans are not only animals of instinct and impulse; humans are endowed with a reasoning power that can calculate how to preserve and secure their lives. Hobbes' laws of nature are the rules of conduct which humans, calculating the means of survival, would observe if they understood their predicament in a universe ruled by passion.

Three of the laws of nature as defined by Hobbes provide the foundation of the social contract. Every human in the state of nature has a natural right to everything, and so no human is physically secure. The first law of nature, therefore, is that every human should seek peace as far as possible, but when peace cannot be obtained, humans may and should resort to war (LN 14). This law of nature really covers two points: the first is that humans should seek peace, and this Hobbes calls the fundamental law of nature; the second is that humans should defend themselves by all the means at their disposal, and this Hobbes calls the sum total of each human's natural rights.

The second law of nature derives from the fundamental law commanding every human to seek peace. As long as every human has a right to do anything he or she desires to do, all humans live in a condition of war. But if all give up their natural rights, that is, their rights to do anything, then there will be peace. The second law of nature, therefore, commands each human to be willing to give up his or her natural right if—but only if—other humans also give up theirs (LN 14). Of course, there is no reason for one human to give up his or her natural rights if others fail to give up theirs. This, says Hobbes, is what the Golden Rule of the Gospel really means when it commands humans to do to others whatever they would have others do to them. Hobbes accordingly restates the Golden Rule in negative form as the law of all human beings: don't do to others what you don't want done to yourself.

The third law of nature, that humans should perform their covenants, derives from the second (LN 15). By the second law, says Hobbes, "we are obliged to transfer to another such rights, as being retained, hinder the peace of mankind." (This is not quite accurate; in the second law, Hobbes said only that humans should be *willing* to transfer their rights to another if everybody else does so.) The third law of nature derives from the second because the failure to keep covenants would mean the condition of war. This third law applies only to mutual promises where there is a civil power to compel the performance of other parties or to the keeping of promises where other parties have already kept theirs.

In what sense do the laws of nature morally oblige individuals? According to Hobbes, "the laws of nature oblige *in foro interno,* that is to say, to a desire that they should take place, but *in foro externo,* that is, to the putting them in act, not always" (LN 15). The reason why humans are not always obliged by the laws of nature to keep their promises is that to do so when and where no one else does so would make those who do perform their promises a prey to others and procure their ruin, "contrary to the ground of all laws of nature, which tend to nature's preservation." Since the basis of all laws of nature is the absolute duty of each human to preserve his or her life, the laws of nature cannot oblige anyone to act according to them if no one else does. But since laws of nature are rational calculations on how to preserve one's life most efficiently, the laws oblige humans to desire to obey them.

The science of the laws of nature is "the true and only moral philosophy," which studies what is good and evil, "names that signify our appetites and aversions" (LN 15). Individuals differ not only in their judgment of what is pleasant or unpleasant but also of what is agreeable or disagreeable to reason. Indeed, the same individual at different times disagrees with himself or herself about what is good or evil. Since individual appetites conflict, there inevitably arise "disputes, controversies, and at last war." So long as humans are in a state of nature, which is a condition of war, "private appetite is the measure of good and evil." The "laws" of nature, therefore, are inaccurately termed laws, "for they are but conclusions or theorems concerning what conduces to the conservation and defense of [individuals] themselves, whereas law properly is the word of him that by right has command over others."

Some question whether there is any morality at all in Hobbes' state of nature.[11] There is certainly no traditional morality. Hobbes recognizes nothing as naturally good for humans except physical survival; Hobbes neither situates humans within a natural order nor credits anything other than physical life as being qualitatively good for humans. Correspondingly, Hobbes assigns no role to reason in morals save that of calculating how best to survive. But Hobbes does recognize a truncated moral order in the state of nature. Humans have a natural inclination to save their lives, and reason, apprehending that, recognizes an absolute moral duty to do so. Secondly, reason can apprehend that the laws of nature are means to achieve the goal of survival and so dictate that humans desire to give up to a sovereign power their natural rights to do anything they desire to do, provided others agree to do likewise. Thirdly, this moral obligation that humans desire to surrender their natural right to do anything they desire to do leads to a commonwealth with the power to insure maximum security and minimum virtue. Thus, the moral obligation in the state of nature to desire peace with, and trust in, others leads to a moral obligation in the commonwealth to act peacefully and trustfully toward others at the command of the sovereign. Hobbes' moral vision, however, in or out of the commonwealth, is limited to the objective of mere physical survival.

Hobbes' theory of the moral obligation of citizen-subjects to execute laws of the commonwealth is as categorical as his theory of the moral obligation of individuals to execute the laws of nature is hypothetical. When individuals execute the second law of nature and covenant with one another to form a commonwealth, the multitude of individuals is united in one person, "that great Leviathan" or "mortal God" (LN 17). And "he that carries this person is called sovereign and said to have sovereign power." Now that a civil power has been created to compel others to keep their promises, the condition for observing the third law of nature has been fulfilled, and so each individual is morally obliged to keep his or her promise to obey the sovereign in all things. There is only one exception to this moral obligation to obey the sovereign. Since individuals yield their natural rights to the sovereign in order to preserve their lives, no individual can be morally obliged to obey the sovereign if the latter commands anything contrary to the individual's right and duty to preserve his or her own life.

In the society formed by the covenanting of citizen-subjects, the sovereign is not merely *a* source but *the* source of moral obligation. No competing claim of moral obligation can be allowed. Accordingly, "it belongs . . . to him that has sovereign power to be judge or constitute all judges of opinions and doctrines as a thing necessary to peace, thereby to prevent discord and civil war" (LN 18). Among the doctrines most dangerous to the peace of the commonwealth, Hobbes includes the view that "every private man is judge of good and evil actions" (LN 19). Such a view Hobbes condemns as seditious. Although every human is the judge of good and evil in the state of nature, "where there are no civil laws, and also under civil government in such cases as are not determined by the law," civil law is otherwise "the measure of good and evil actions." If every human were to be judge of good and evil in such cases as civil law has determined, then humans would be "disposed to debate with themselves and dispute the commands of the commonwealth and afterwards to obey or disobey them as in their private judgments they shall think fit," and thereby "the commonwealth is distracted and weakened."

Similarly, "another doctrine repugnant to civil society is that whatsoever a man does against his conscience is sin" (LN 19). Such a doctrine "depends on the presumption of [every man] making himself judge of good and evil." But "the law is the public conscience, by which he has already undertaken to be guided." If the moral claims of private conscience for each individual were superior to those of public conscience, the commonwealth would be distracted by a diversity of moral judgments, and no human would obey the sovereign power "further than it shall seem good in his own eyes."

According to Hobbes sin is "nothing but the transgression of the law," and spiritual authorities can claim no right independent of the sovereign to declare what is law" (LN 29). As for the possibility that divine law runs counter to civil law, Hobbes claims that "a subject that has no certain and assured revelation particularly to himself concerning the will of God is to obey for such the command of the commonwealth" (LN 26). Hobbes argues further that "if man were at liberty to take for God's commandments their own dreams and fancies . . . scarce two men would agree upon what is God's commandment, and . . . every man would despise the commandments of the commonwealth." Hobbes concludes, therefore, "that in all things not contrary to the moral law (that is to say, to the law of nature), all subjects are bound to obey that for divine law which is declared to be so by the laws of the commonwealth."

Hobbes admits that it is "equity . . . that every man equally enjoy his liberty" in everything not regulated by the commonwealth (LN 26). Thus, the liberty of subjects lies "only in those things . . . the sovereign has praetermitted [omitted to regulate]" (LN 21). Hobbes advises the sovereign to allow "the liberty to buy and sell and otherwise contract with one another," the liberties of subjects "to choose their own abode, their own diet, their own trade of life," and the liberty to "institute their children as they themselves think fit." In short, the sovereign should follow a policy of laissez-faire as far as compatible with the purpose of the commonwealth, namely, the physical survival of subjects. But liberty for Hobbes is simply the "absence of external impediments, which impediments may oft take away part of a man's power to do what he would but cannot hinder him from using the power left to him according as his judgment and reason shall dictate to him" (LN 14). And since sin is "nothing but transgression of law," every man is free of any moral obligation to do anything other than what he desires to do with respect to those things that the sovereign allows.

This policy of laissez-faire poses an obvious problem for the Hobbesian commonwealth. In the absence of legal prescription by the sovereign, individuals have no moral obligation except to preserve their own lives, and individuals are thus effectively in a state of nature with respect to those things that the sovereign praetermits. True, the laws of nature morally oblige individuals not to harm others on the proviso that others do not harm them. But the state of nature is a state of unrestricted appetite. A legal policy of economic and personal laissez-faire would seem inevitably, therefore, to encourage conflict and so undermine the security of the commonwealth. Hobbes can hardly have wanted that result, and he may have sought to avoid it by enrolling the Christian religion in the service of the commonwealth. Religion, under the sovereign's control, can supplement the civil laws by providing subjects with other-worldly motives for and principles of self-restraint. In this way, civic religion helps to accomplish indirectly what the sovereign would otherwise have to accomplish entirely by force, and so a legal policy of economic and personal laissez-faire may become compatible with the peace of the commonwealth.

Civic relation is an integral part of the Hobbesian commonwealth. "Reason directs [humans] not only to worship God in secret but also and especially in public" (LN 31). Since the commonwealth is but one person, the commonwealth "ought to exhibit to God but one worship, which then it does when it commands it to be exhibited by private man publicly." And this public worship is to be uniform since "those actions that are done differently by different men cannot be said to be a public worship." Hobbes is totally Erastian in his subordination of the church to the state. A church is defined as "a company of men professing Christian religion united in the person of one sovereign, at whose command they ought to assemble, and without whose authority they ought not to assemble" (LN 39). Moreover, as previously indicated, the sovereign determines what doctrines are fit to be taught. Thus, the sovereign entirely decides the religious doctrine and practice of the established church, and the established church functions only as an instrument of the sovereign's governance of the commonwealth.

Locke

Like Hobbes, Locke situates individual humans initially in a prepolitical state of nature. In this natural condition, humans have perfect freedom "to order their actions and dispose of their possessions and persons as they think fit, within the bounds of the law of nature, without asking leave or depending upon the will of any other man" (TG 2. 4). But though the state of nature is a state of liberty, it is not a state of license. "The state of nature has a law of nature to govern it which obliges everyone" (TG 2. 6). Not only do humans have no liberty under the law of nature to destroy themselves or their own possessions, except for use, but also "reason, which is that law, teaches all mankind who will but consult it that, being all equal and independent, no one ought to harm another in his life, health, liberty, or possessions." All humans are "the workmanship of one omnipotent and infinitely wise Maker," all His servants, all sent into the world by His orders and for His business, and so "they are His property whose workmanship they are, made to last during His, not one another's pleasure." As everyone is bound to preserve himself, so ought he not impair the life, liberties, and properties of another unless necessary to preserve his own life, liberties, and properties. Even when using force to protect their own rights, humans have no arbitrary power over the criminal but "only to retribute to him, as far as calm reason and conscience dictate, what is proportionate to his transgression" (TG 2. 8). Moreover, "truth and keeping of faith belong to men as men and not as members of society" (TG 2. 14).

Thus, while individuals have no natural inclination or corresponding moral obligation to live in organized society, that is, to cooperate with others toward a common good, Locke does acknowledge in the state of nature certain minimal moral obligations to others. The laws of nature seem to mean something quite different for Locke than for Hobbes. For Hobbes, the laws of nature mean only rules of reason calculated to help preserve the individual's life and insure his or her physical survival; for Locke, however, the laws of nature seem to include elements of the traditional natural law such as the Golden Rule.

If Locke is consistent with his empirical epistemology, he cannot mean to say that reason apprehends real essences or natures as the basis of the law of nature. Of course, Locke may be claiming for reason only that it can apprehend the practical utility of respecting the rights of others and of keeping one's promises to others for one's own security. In fact, Locke himself cites Hooker, an English philosopher-theologian of the previous century, in the same chapter to the effect that "if I do harm, I must look to suffer, there being no reason that others should show greater measure of love to me than they have by me showed unto them" (TG 2. 5). If this interpretation of Locke's law of nature is correct, then Locke would differ from Hobbes only on the policy that reason calculates to be useful for survival and security in the state of nature.

But Locke means more by the laws of nature than merely pragmatic rules of reason for physical survival and security; over and above pragmatic reasons, the lives, liberties, and possessions of others are to be respected because God, in creating men equal, has willed it so. There is here a vestige of the traditional natural law doctrine, although appeal to divine law is warranted in Locke's philosophy only on fideistic, not rational, grounds. Locke may be arguing, in effect, that the laws of nature are in accord with *both* narrowly pragmatic reason *and* a natural religious faith.[12]

However reasonable and religious the laws of nature, the state of nature is, in fact, a state of war similar to the Hobbesian model. Since the state of nature is a condition of affairs where there is no common authority, God puts man "under strong obligations of necessity, convenience, and inclination to drive him into society" (TG 7. 77). In the state of nature, man's enjoyment of his rights to be absolute lord of his own person and possessions is "very uncertain and constantly exposed to the invasion of others." All "being kings . . . and the greater part no strict observers of equity and justice, the enjoyment of the property he has in this state is very unsafe, very unsecure" (TG 9. 123).

Locke enumerates three main things as wanting in the state of nature. First, "men, being biased by their interest as well as ignorant for want of studying it [the law of nature], are not apt to allow of it as a law binding to them in the application of it to their particular cases" (TG 9. 124). Wherefore, it is desirable that there should be "an established, settled, known law, received and allowed by common consent to be the standard of right and wrong and the common measure to decide all controversies between them." Second, in the judgment and execution of the law of nature, "men being partial to themselves, passion and revenge is very apt to carry them too far and with too much heat in their own cases, as well as negligence and unconcernedness to make them too remiss in other men's" (TG 9. 125). Wherefore, it is desirable that there be "a known and indifferent judge with authority to determine all differences according to the established law." Third, in the state of nature, "there often wants power to back and support the sentence when right and to give it due execution" (TG 9. 126). Thus, "notwithstanding all the privileges of the state of nature, being but in an ill condition while they remain in it," humans are "quickly driven into society" (TG 9. 127).

This state of war makes individuals "willing to quit a condition which, however free, is full of fears and continual dangers" and "to join in society with others . . . for the mutual preservation of their lives, liberties, and estates" (TG 9. 123). When individuals form the commonwealth, they give up their two principal powers in the state of nature. First, they give up the power of each to do whatever he thinks fit for the preservation of himself and others. The law of nature by itself is insufficient to secure for the individual his life, liberties, and estates because of "the corruption and viciousness of degenerate men" (TG 9. 128). When individuals unite into an organized society, they give up their power to do whatever they think fit to secure their lives and possessions, and so the laws of society restrict the liberty that individuals had by the law of nature. Second, individuals give up the power on their own authority to punish violations of the law of nature (TG 9. 128). In the commonwealth, individuals commit their natural forces "to assist the executive power of the society as the law thereof shall require" (TG 9. 129).

When individuals consent to unite into society, the majority has "the whole power of the community" and "may employ all that power in making laws for the community" (TG 10. 132). Since the purpose of human beings uniting into society is to secure the safety of their lives and properties, and, since the laws of society are the means thereto, "the first and fundamental positive law of all commonwealths is the establishing of the legislative power" (TG 11. 134). This legislative power is not only the supreme power of the commonwealth but "sacred and unalterable in the hands where the community have once placed it." Therefore, "all the obedience which by the most solemn ties anyone can be obliged to pay ultimately terminates in this supreme power and is directed by those laws which it enacts."

But the legislative power, though supreme, is not "absolutely arbitrary over the lives and fortunes of the people" (TG 11. 135). First, the legislative power is limited by the law of nature for the public good of society, that is, the preservation of each person's life and possessions. The laws that legislators make should conform to the fundamental law of nature for the preservation of human lives and estates, and so no human law can be valid that operates against the law of nature. Second, laws should be established, promulgated, and applied equally to all (TG 11. 136). Third, taxes should not be levied without the consent of the people, given by themselves or their deputies (TG 11. 138). Fourth, the legislative power cannot be delegated to anybody else (TG 11. 141).

This brief survey of Locke's construction of civil or politically organized society shows both the kinship of his views to those of Hobbes and the differences between their common views and those of premodern political theory. Both Hobbes and Locke deny that humans have any natural inclination to cooperate with others toward common goals. Both rationally ground social organization on the desire to preserve one's life and, in the case of Locke, also one's possessions.[13] Both deny that organized society has any scientific, cultural, or moral goal. Men unite in society only to secure peace, not to seek truth, cultivate beauty, or strive for virtue.

But Locke appears to differ from Hobbes on the limits of sovereign power in the commonwealth as a matter of principle, although Hobbes would advise the sovereign to limit his exercise of power as a matter of policy. Three of the four limitations Locke imposes on rightful exercise of legislative power are procedural. Procedural norms are no small virtue, but they are, after all, only procedural. The other normative limit on the exercise of sovereign power seems to be substantive; the law of nature itself limits the legislative power to the preservation of each human's lives and possessions. Accordingly, Locke declares that the legislative power "can never have a right to destroy, enslave, or designedly to impoverish the subjects" (TG 11. 135). Yet, there is less substantive limit on legislative power here than meets the eye. Humans unite into society precisely because humans are poor judges in their own cases and need "a known and indifferent judge with authority to determine all differences according to the established law," and the legislative power, as supreme, alone judges what is necessary to preserve men's lives and possessions. An individual or minority has no rights against the judgment of the majority, and there is no objective order of nature to appeal to against the judgment of the legislative majority. Locke in principle disavows only the right of a legislative majority or its deputies to destroy, enslave, or impoverish subjects by *design.*

Second, how are the rights of individuals to their lives and possessions more secure in the commonwealth from deprivation by the majority under the law of nature than they were in the state of nature from deprivation by other individuals? Locke has argued that the law of nature alone is insufficient to secure to the individual his life, liberties, and estates because of "the corruption and viciousness of degenerate men." But when the very same corrupt and vicious men unite into society, they will still be disposed to invade the rights of others. In the commonwealth, of course, the sovereign power will protect individuals from deprivations by other individuals, but what will restrain the majority from depriving individuals or minorities of their rights? Locke does not specifically address the question, in part because his practical concern is to vindicate the rights of Parliament against Stuart claims to absolute royal power. The question, however, remains a serious one, and other elements of Locke's political theory suggest several answers.

As indicated above, Locke prescribes three procedural limits for the exercise of legislative power: (1) laws should be properly established, duly promulgated, and applied equally to all; (2) taxes should not be levied without the consent of the people or the people's deputies; and (3) legislative power cannot be delegated. In addition, the British Constitution after 1688 provided Locke with a model of legislative power. Under that constitution, Kings, Lords, and Commons all participated in the legislative power, and laws could be enacted only with the assent of all three units. The requirement that the three units concur in the exercise of legislative power would serve to check tyranny against individual rights by any one unit. And Locke would undoubtedly accept Madison's argument in Federalist #10 that diversity of interests within the Houses of Parliament would make the formation of a tyrannical majority less likely.

But conceding the advantages of procedural and institutional arrangements to deter tyranny, a considerable advance over medieval political practice, the problem of morally virtuous citizens remains. In the final analysis, morally virtuous citizens are a necessary condition for virtuous government of the commonwealth. If the mass of citizens is "degenerate," to use Locke's phrase, then the best of procedural and institutional arrangements cannot save the commonwealth from tyranny in the long run. Hobbes may have recognized the importance of religion for this purpose insofar as he made religion an instrument of the sovereign's power. Locke, too, makes a place for religion in the commonwealth, but he divorces the private business of religion from the public business of civil government.

Locke declares in his *Letter Concerning Toleration* that it is "above all things necessary to distinguish exactly the business of civil government from that of religion and to settle the just bounds that lie between the one and the other" (JL 5:9). The commonwealth, says Locke, is "a society of men constituted only for the procuring, preserving, and advancing their own civil interests" (JL 5:10). All civil power is confined to the care of promoting "life, liberty, . . . indolency of body, and the possession of outward things," and "it neither can nor ought to be extended to the salvation of souls." Among other reasons, Locke argues that "the care of souls cannot belong to the civil magistrate because his power consists only in outward force," while "true and saving religion consists in the inward persuasion of the mind" (JL 5:11). Thus, "all power of civil government relates only to men's civil interests, is confined to the care of things of this world, and has nothing to do with the world to come" (JL 5:13). Unlike Hobbes, Locke recognizes a moral obligation on the part of individuals to follow their religious conscience without compulsion or restraint from the civil power because "no way whatsoever that I shall walk in against the dictates of my conscience will ever bring me to the mansions of the blessed" (JL 5:28).

"Not the least part of religion and true piety" consists in a good life, and a good life "concerns also the civil government" (JL 5:41). The good life relates to the safety "both of men's souls and of the commonwealth," and so "moral actions belong . . . to the jurisdiction both of the outward and the inward court, . . . both of the magistrate and conscience." Locke acknowledges the potential for conflict if the magistrate enjoins something the individual conscience judges to be morally objectionable. While he deems that faithful administration of government makes such a case unlikely, if it should occur, "a private person is to abstain from the actions he judges unlawful," but he is also "to undergo the punishment, which is not unlawful for him to bear" (JL 5:43). Individuals, though morally obliged to disobey commands against conscience, should accept punishment therefor because "the private judgment of any person concerning a law enacted in political matters for the public good does not take away the obligation of that law nor deserve a dispensation.''

Locke is thus circumspect on the rights of conscientious objectors to the commands of governments. On the one hand, an individual who judges a command of government to be against his or her religious conviction is morally justified, even morally obliged, to refuse to perform the command; on the other hand, an individual is morally obliged to submit to punishment for his or her disobedience. The individual may be physically obliged, of course, to submit to such punishment (although flight or concealment may be options), but why should the individual be considered *morally* obliged to submit to punishment for not doing what the individual was in fact morally obliged not to do?

Locke does not explain here in detail why individuals morally obliged to disobey magistrates' commands are morally obliged to accept punishment for their disobedience, but Locke's general political theory makes the reason clear. When individuals unite into society, they give up their powers, that is, their rights, to do whatever they think fit for their security and to judge what is fit to procure that security. *All* laws intended for the public good are morally obligatory on individuals, and only the legislative power can judge whether or not actions are suitable for the public good. Locke differs from Hobbes only in the way in which laws morally oblige individuals in cases where religious beliefs conflict with the laws. For both, laws of the commonwealth morally oblige individuals simply because individuals by their own consent upon uniting into society give up their rights to do and to judge as they think fit for their own security. For Hobbes, all laws morally oblige individuals maximally to performance of the acts commanded (save acts contrary to self-preservation); for Locke, laws commanding acts contrary to religious beliefs morally oblige individuals minimally to acceptance of punishment.

All this differs from medieval political theorists like Aquinas, for whom reason is capable of apprehending an objective order of nature as the point of reference to determine the moral responsibilities of individuals to governments and of governments to individuals. Aquinas acknowledges both that individuals have a moral obligation to disobey laws commanding actions that individuals, correctly or incorrectly, regard as contrary to divine or natural law, and that individuals have no automatic moral obligation to submit to punishment for their morally obligatory disobedience. Individuals may have a moral obligation to accept punishment for

extrinsic reasons, for example, to avoid scandal or civil distur-
bance, but not simply because they disobey a command of civil
authorities. At the same time, when individuals misinterpret divine or natural law, governments may be justified by the common
good in inflicting punishment for disobedience as they were in
issuing the initial command.

Hobbes' perspective on private virtue and public duty is radically different from that of Aristotle and Aquinas. Private virtue in
the sense of individual behavior conforming to the demands of a
rationally apprehended order of moral goods beyond self-preservation is eliminated. Each human has a natural right to do anything
he or she desires to do. As a result, the state of nature is a state of
war. The Leviathan commonwealth rescues humans from themselves. Calculating what is necessary for survival, individuals consent together to surrender their natural rights (save the right of
self-preservation) to the sovereign. And so public duty becomes not
only *a* moral obligation but *the* moral obligation; private virtue
and public duty coalesce into one. Subjects are good if and only if
they obey the sovereign, and the sovereign is good if and only if he
protects the lives of his subjects. True, Hobbes advises the sovereign to pursue a broad laissez faire policy, but that is merely
advice.

At first blush, Locke appears to differ substantially from
Hobbes on both private virtue and public duty. Locke admits
elements of an objective moral order based on the will of God (e.g.,
the Golden Rule). But this order is accessible only through a
natural disposition toward faith in God. Reason cannot apprehend the moral good because it cannot apprehend an order of
nature. So reason is for Locke, as for Hobbes, a calculating reason,
not an understanding reason, even if calculating reason comes to a
more sociable calculation for Locke than it does for Hobbes.

Despite this minimal moral order in the state of nature, the
viciousness of humans turns the state of nature into a state of war.
For Hobbes as well as for Locke, the commonwealth is necessary to
rescue humans from themselves. Individuals, calculating how to
preserve their lives and their possessions, consent together to give
up to the legislative power of the commonwealth their natural
rights to do and judge what they think best for their security. They
thus morally oblige themselves to obey the will of the majority.
Although the legislative power is limited to procuring the public
good, that is, the preservation of human lives and estates, only the
legislative power is the judge of what is suitable to procure that

good. Unlike Hobbes, Locke recognizes moral obligations of individuals beyond those to the commonwealth. Like Hobbes, however, public duty is the ultimate arbiter of morality in the external forum. However much Locke recognizes the moral obligation of individuals to disobey laws commanding actions against their conscience, he insists that disobedient individuals have a moral obligation to accept punishment therefor.

Hobbes and Locke recognized the potential of egotistical individuals for conflict and the necessity of legitimated force to control it. Not that Hobbes and Locke were unique in this respect. Medieval political theorists like Aquinas also recognized that men act egotistically against reason, and that governments are necessary to coerce deviants into obedience. Medieval theorists, linking political obligation to an objective order of justice, relied on the reason of rulers and citizens to distinguish just from unjust laws. But despite the ideal of justice in medieval political theory, medieval rulers and citizens in practice disputed endlessly about what was just or unjust, and modern religious and moral pluralism has compounded the problem. Hobbes and Locke are at least unambiguous on how to solve the problem: Hobbes would morally oblige individuals to obey all commands of the sovereign, save those contrary to self-preservation, and Locke would morally oblige individuals to obey all commands of the legislative power or at least to accept punishment for disobedience of the legislative power. Moreover, Locke specified the legislative power as the will of the majority operating under a constitution.

But the analyses of Hobbes and Locke are inadequate for several reasons. First, Hobbes ultimately subordinates individuals to the commonwealth in every respect save that of self-preservation. And Locke ultimately does the same because the legislative power of the commonwealth alone can judge what is suitable for survival and security. Hobbes denies individual rights utterly, and Locke, in the final analysis, consigns them to the mercy of the majority.

Second, if human reason cannot apprehend moral good, and human will is under no constraint of reason through apprehension of moral good, individuals are not likely to follow the argument of Hobbes and Locke that they are morally obliged to obey the sovereign or legislative power simply because they actually or putatively promised to do so. Hobbes and Locke may be right that political obedience is in the long-term interest of individuals for physical survival and security, but passionate man may well prefer short-term interests. As Lord Keynes quipped to Winston Churchill, everybody is dead in the long run. How is passionate man to be disciplined to pursue long-term interests? Hobbes may have recognized this problem by resort to a state religion, and Locke may have done so by resort to private religion. In any case, both relied, at least tacitly, on the support of biblical religion to restrain individual appetite and so make the social and political enterprise viable.

Lastly, the "realism" of Hobbes and Locke is not realistic enough. Both Hobbes and Locke found organized society on the need to control violence, and control of conflict is certainly an essential reason why humans organize into societies. But there is another essential reason why humans do so; humans as rational persons naturally incline to cooperate with their fellows to achieve long-term common goals even at short-term costs to individuals.

Marx

At the other extreme from the individualism of Hobbes and Locke is the collectivism of Karl Marx, as interpreted by Nicolai Lenin. It is appropriate to couple the philosophy of Lenin with that of Marx because Lenin, after the October Russian Revolution, put Marxist theory into practice for the first time and thereby provided a sort of laboratory model revealing the implications of that theory. Our concerns here center on communist perspectives on individuals, society, and state, although those perspectives cannot be studied apart from economic theory.

Marx was a dialectical materialist who conceived economics as the decisive determinant of human history. Everything in human history revolves around the dominant modes of production. In the course of human history, there have been a succession of economic systems:[14] slavery, feudalism, and capitalism. Each system is distinguished by the class which controls the means of production and the method by which the means of production are controlled. Each system contains within itself the seeds of its own destruction, which is inevitable. Slavery collapsed because slaves exercised no control over what they produced and so became alienated from themselves as producers and from the organization of slave society. The dissatisfaction of the slave class mounted to a critical point, and the economic system of slavery collapsed. The process was then repeated in the successor system, feudalism. The bourgeois classes of moneylenders and merchants created by feudal lords

exercised no control over what they produced and so became alienated from themselves as producers and from the organization of
feudal society. When the dissatisfaction of the bourgeois classes
reached the critical point, the economic system of feudalism collapsed. The process will be repeated one more time. The working
class created by bourgeois entrepreneurs exercises no control over
what it produces, and so workers become alienated from themselves as producers and from the organization of bourgeois society.
When the dissatisfaction of the working class reaches the critical
point, the economic system of capitalism will be overthrown.

Marx was not a crass materialist. He does not deny the existence
of such things as ideas, ideals, religion, culture, and art. But Marx
insists that these are superstructures erected on an economic foundation, by-products of an economic system. The dominant class in
each economic system adopts whatever ideas and cultural patterns
that will reinforce the dominance of that class. Religion, law,
mores, and civic culture are inventions by the dominant class to
perpetuate the system. So too is political power a superstructure
built to maintain an economic system; the dominant class uses the
machinery of state to maintain economic power and to repress the
working class.

Marx thought capitalism intrinsically unjust not only because
it excluded workers from control over what they produced, but also
because it robbed workers of the fruits of their labor. To understand why Marx condemned capitalism as expropriation, it is
necessary to understand his theory of value. All products of utility
derive their value exclusively from the labor that goes into their
production. Capitalists, however, pay workers wages equal to only
part of the value of the workers' labor. The difference or surplus
value from sales of the product goes to the employer. Thus, the
profits of employers are nothing but the value of workers' labor
above the wages they are paid.

The interests of the employer class and the working class are
irreconcilable. Since one class benefits at the expense of the other,
the two have no common interest. The dialectic of history will
increase the conflict between the classes, and the dynamics of
capitalism will bring about its own destruction. The capitalist
system of production, the factory system, initiates a process of
socialization when workers in factories interact with one another
to produce manufactured goods. Then, as a result of capitalist
competition, less efficient smaller employers are eliminated, and
production becomes centralized in larger and larger factories with

still greater socialization. But at the same time, there is a corresponding increase in the level of poverty and exploitation; capitalists hold down workers' wages in order to compete successfully, and fewer and fewer capitalists will survive the competition. Thus, the capitalist system, by its own raison d'être, tends inexorably to increase both the numbers of the working class and that class's misery. Eventually, a worldwide depression results when workers are unable to purchase the goods they produce. Since there will be too few capitalists at this end-point to control the wrath of the working class, the capitalist system will be overthrown, and the expropriators expropriated.

In order to overturn the capitalist economic system, it is necessary for the working class to get control of the machinery of state. Marx himself is ambiguous about how this is to be achieved. He typically claims that workers will need to resort to revolution to gain control of the state, but occasionally he speaks of winning the struggle by democratic means. Lenin interpreted Marx as a revolutionary and attempted to justify the October Russian Revolution against the democratic Kerensky government in Marxist terms. At least from that date, Marxism and communism have been linked to violent revolutions against democratic and other noncommunist systems.

After the revolution overthrows the capitalist political structure, the working class or proletariat will assume a dictatorship to reconstruct society without classes. The working class will need to eradicate the institutional and cultural vestiges of bourgeois society. The capitalist system, in control of the means of production for centuries, has erected a vast superstructure to maintain that control, and so the working class, when it comes to power, must use the machinery of state to dismantle the superstructure. The working class dictatorship will control and reconstruct not only the economy but also every facet of culture: family, education, art, and religion.

The assumption of political power by the working class does not mean democracy as we understand the term. Bourgeois ways of thinking will survive the bourgeoisie's loss of political power, and so an elite will be needed to supervise the reconstruction of society. That elite is the Communist party, the faithful followers of Marx. Lenin intended, however, that there should be a democracy of sorts within the party. Party members at all levels should participate in open discussion about major policy decisions, but they should carry out the decisions reached without any further question. This mixture of limited democracy and centralized authority Lenin called "democratic centralism."

The dialectic of history comes to completion in a reconstructed society without any class divisions.[15] With the elimination of the capitalist class, control of the means of production will have been returned to workers, and the source of alienation and conflict removed; individuals will now be at peace with themselves and with one another. With the reconstruction of social consciousness under the dictatorship of the working class, that is, the Communist party, individuals will recognize that their private interests are perfectly reconciled in community control of the means of production. Individuals will recognize that they should contribute to the economic product according to their abilities, and that they should receive from that product according to their needs. The state will no longer be necessary in a world made up solely of workers conscious of their solidarity. The state will simply wither away.

From the Thomistic perspective, the Marxist view radically misconceives the nature of human personality. For the Marxist, the role of economic producer is the core of human personality, and all other roles revolve around it. For the Thomist, rational activity is what makes human persons specifically such, and all other activities are subordinate to it. Put another way, Marx mistakenly identified economic good with human good.

Marx did not originate the concept of *homo economicus* (economic man). That had already been done before Marx by English laissez-faire economists. Those economists advocated unfettered free enterprise. Leave the entrepreneur free to maximize his profit, and the wealth of nations would grow as if by an invisible hand. Production will become more efficient in the face of competition, inefficient producers will be eliminated, and consumers will be the beneficiaries. Entrepreneurs will invest their profits in new productive units, and so employment will expand.

Laissez-faire theory presupposes that the supreme social good is the wealth of nations, and that the chief function of reason is to calculate how to increase economic growth. In this scheme of things, the short-term consequences for individual workers and their families are irrelevant so long as the collective wealth of the body politic increases. Thus, the long-term economic good of the collective whole was absolutized over the distributive economic and social consequences for individuals and families. The free enterprise world is basically an amoral jungle in which entrepreneurs war against other entrepreneurs without regarding the consequences for workers, at least in the short run. Laissez-faire theory enshrined Hobbes' state of nature in the economic sphere under the eye of a benign sovereign.

Marx took over that view of human personality. He too held that aggregate economic good was human good, but he stood the classical economists' theory on its head. Workers could play the same game as entrepreneurs. If workers became conscious of their exploitation by entrepreneurs, they could organize themselves to take over the means of production from the entrepreneurs.[16] Thomas P. Neill aptly calls Marx's worker nothing more than the classical economists' economic man "with dirty hands."[17] Marx would have workers pursue their economic interests as ruthlessly as the classical economists would have entrepreneurs pursue theirs. Neither the classical economists nor Marx allow their respective prototypes the capacity to be human in the fullest sense, with economic good subordinate to the higher goods proper to the human spirit.

Marx shared with the classical economists the view that economic good was the archetechtonic good of human personality. (One might add that the modern welfare state seems to share much the same view.) Marx, to his credit, added an element of justice into his calculus of economic good. Marx recognized that the long-term aggregate economic good of the body politic was not enough to satisfy justice, and the short-term economic welfare of individual workers and their families should be taken into consideration. Yet economic justice is only one of the virtues. Justice itself is more inclusive than economic justice, moral virtue more inclusive than justice, and there are intellectual as well as moral virtues.

From Marx's identification of human good with economic good flows his conception of the relationship between human persons and organized society.[18] Marx would have human persons be simply parts of a collective whole and allow the parts to have no proper good apart from the good of the whole. Marx thereby makes animal society the analogue for human society. As, for example, individual bees are only economic producers for the good of the hive, so individual humans are only economic producers for the good of human society; neither individual bees nor individual humans have a proper good apart from the good of the whole.

But because the human person is more than an economic producer, human society has other and higher purposes than economic production. Humans have the capacity and the desire to inquire after truth, to appreciate beauty, and to love other humans.

This spiritual dimension is at the core of human personality and can only be achieved in concert with others. Organized society, therefore, exists to provide the human person not only with conditions of material well-being but also and especially with conditions conducive to the communication of truth, beauty, and goodness.

The notions of the good of the person, the good of the social part, and of the common good, the good of the social whole, are correlates. The common good is the good of persons. On the one hand, this means that the human good is not the aggregate of the goods of individual members of society. That was the concept of nineteenth-century laissez-faire economists, and the concept is basically anarchistic since it denies the existence of any proper good of the whole. On the other hand, the common good, as the good of human persons, is the good of the members of society. The common good is a good common *both* to the whole *and* to the parts, a good which should redound to the benefit of the parts. To be true to its own finality, the common good requires recognition of the rights of individual human persons, families, and associations. In short, the maximum freedom of the human persons who constitute society is part of the common good and indeed its principal part.

The finality of the common good requires that the organization of society respect what surpasses it, namely, the sphere of the human spirit: truth, beauty, goodness, friendship, and love. The common good of society is ordered to the spiritual goods of human personality and not the other way around. In fact, it is a principal part of the common good to foster the spiritual goods of human personality. From this perspective, the proper concept of organized human society is neither purely individualist nor purely communal. Properly conceived, there is no opposition between the good of the individual human person and the good of the community. Rather, the two goods are reciprocally subordinated to each other in different respects and mutually imply one another when the relation of both to human personality is rightly understood.

Both laissez-faire economists and Marx polarized individual and community. In that "either-or" dichotomy, the laissez-faire economists opted for individual good, and Marx opted for collective good. Aquinas chose a middle ground, refusing to submerge either good totally under the other. "Each man, in all that he is or has, belongs to the community, just as a part, in all that it is, belongs to the whole" (ST 1-2. 96. 4; cf. 2-2. 61. 1, 64. 2, and 65.1).

Human persons are members of society and so ordered in all that they are and have to the common good of society. On the other hand, "man is not ordered to the body politic *according to all that he is and has* [italics added]" (ST 1-2. 21. 4. ad 3). That is to say, there are some things about human personality that transcend political society and are not subject to determination by that society.

We may illustrate this paradoxical distinction between person-as-part and person-as-whole with an example. Scientists depend on society for their intellectual environment and development: society provides the schools, libraries, laboratories, publications, and so forth, that are necessary for scientific inquiry. Conversely, society can call scientists to its service as soldiers in time of war. From this perspective, scientists, in all that they are and have, are ordered to the common good of society. But scientific (or any other) truth in essence transcends the temporal order of the political community, does not depend on that community, and cannot be subject to the command of the community. And so no community or its instrumental agency, the state, has any right or competence to require a scientist to hold or teach one scientific theory rather than another because the theory is more compatible with the ideology or mores of the community.

There is an obvious tension between the dual roles of person as part of the social whole and of person as an autonomous whole. It is here, if anywhere, that a sort of historical dialectic can be said to be at work. For most of human history, the organization of the social whole did not much respect the personality of its individual parts. The evolution of Western democratic freedoms redressed that imbalance and recognized the rights of persons as wholes. But persons living in democratically organized societies may, in turn, lost sight of their essential relationship to the social whole and so fail to fulfill their role as parts of that whole.

Let us assume for the moment that Marxist-Leninist theory is correct about the inherent and unqualified injustice of capitalism. (Marx had ample reason to condemn many features of nineteenth-century capitalism, and so have we to condemn features of twentieth-century capitalism.) From that assumption, it might follow that socialism should replace capitalism as an economic system. But it would not follow that a dictatorship of the proletariat is necessary or proper to reconstruct the economic system, unless one also assumes that human persons have no capacity to reason about their world and to decide on that basis how it is to be economically organized. Marxist-Leninists claim that their economic theory is objective. If so, then it should be accessible to human reason, communicable, and capable of implementation by democratic political processes.

Marxist-Leninists would justify the dictatorship of the proletariat on the ground that bourgeois culture over the centuries has so brainwashed the masses that democratic processes cannot be relied on to bring about the new economic order. To those who do not share the Marxist-Leninist faith, that justification will seem rationalization. Culture undoubtedly conditions the acceptability of ideas, but not totally. Marx and Lenin themselves arrived at their communist world view without any help from a dictatorship of the proletariat; presumably other humans can also do so in the free marketplace of ideas. Human thought is ordered to truth, and truth is communicable. As a theory, Marxism can be debated rationally. But Marxism-Leninism represents more than a secular economic or political theory; it represents a quasi-religious creed. Despite its protestations of objectivity and rationality, Marxism-Leninism is ultimately a matter of belief, not reason. Marxist-Leninists will allow neither the premises nor the conclusions of their theory to be submitted to rational analysis.

Marxist-Leninist theory tells us that when society has been so reconstructed that all private interests are perfectly reconciled into a collective whole, the state will wither away. It is now almost 70 years since the October Revolution, and the Soviet dictatorship of the proletariat in Russia has not yet withered away. Conceding the necessity of maintaining military forces to ward off enemy capitalist societies, one would expect the necessity for internal domestic force at least to have begun to wither away. That will not sway the true Marxist-Leninist believer, but it should give the rest of us pause. Could it be that the idea of freedom is not simply a bourgeois idea? The recalcitrance of many people in the Soviet Union to reconstruct their thinking over this period of time could be seen as confirming the deep-seated human desire to be persons-as-wholes as well as persons-as-parts.

If, from Thomistic perspectives, Marxism-Leninism overuses the machinery of state to reconstruct society under the dictatorship of the proletariat, it seems to underestimate the need of the state after society has been reconstructed along socialist lines. According to the theory, members of a classless society will be perfectly at peace with one another and society. All will contribute whole-heartedly to economic production according to their abilities and be completely satisfied to consume according to their needs.[19] But what workers "need" depends to some extent on what they think they need, and it is unlikely that all will have the same view on that subject. Since material resources are limited and will be as far into the future as the human mind can conceive, allocations of those resources will be necessary. Collective decisions will have to be made about what is produced, how much of each item will be produced, and conditions of work. Those decisions will be authoritative decisions even if they require no physical coercion of workers to be put into effect in a reconstituted society. In that sense, the state cannot wither away. One may choose not to call the machinery of authoritative decision making for society "the state," but that is what it really is.

Nor is it easy to see why the removal of economic causes of deviant social behavior will eliminate other causes of such behavior. Homicides, for example, are typically the product of family quarrels or quarrels between close acquaintances, in which human passion simply overwhelms human reason without any obvious connection with economics. Most mental illness, which may lead to antisocial behavior, appears to have little or nothing to do with economics. Ethnic antagonisms seem more tribally than economically based. Because there are other than economic causes of deviant social behavior, physical coercion would be necessary even in an economically and socially reconstructed society.

Marxism-Leninism has tremendous appeal to many minds, especially in the Third World. That is because Marxism-Leninism is partially and fundamentally correct when it affirms the twin values of economic justice and human solidarity. Democratic societies cannot hope to blunt that appeal by citing the consequences of Marxism-Leninism for human freedom without regard to issues of economic justice; for those who are hungry, economic justice will always seem more important than human freedom. Only if democratic societies commit themselves wholeheartedly to human solidarity and more equitable worldwide economic distribution can they effectively undercut the appeal of Marxism-Leninism.

Notes

1. Aquinas' own work, *De regno, ad regem Cypri*, was melded in the first quarter of the fourteenth century to another work by a disciple of Aquinas, Tolomeo of Lucca. I cite the reconstructed text of Thomas, *On Kingship, to the King of Cyprus*, translated by Gerald B. Phelan and revised by I. Th. Eschmann

(Toronto: Pontifical Institute of Medieval Studies, 1949). On the date of the work, see the introduction to the Phelan-Eschmann text, pp. xxviii–xxx.

2. The source of the teaching here is Avicenna *De anima* 5. 1.

3. I summarize here the excellent development of the distinction between body politic and state by Jacques Maritain, *Man and the State* (Chicago: University of Chicago Press, 1951), pp. 9–19.

4. Cf. Aristotle *Politics* 3. 7. Aristotle uses the word "polity" in both the general sense of "constitution" and, as here, in the special sense of "constitutional democracy."

5. The translation is that of the English Dominican Fathers. Here I have followed the revision of that translation by Dino Bigongiardi, ed., *The Political Ideas of St. Thomas Aquinas* (New York: Hafner, 1953), p. 88; the Bigongiardi revision is more faithful to the Latin text. On the relation of the polity described in this passage in the *Summa* to the "tempering" alluded to in the treatise *On Kingship*, compare R.W. and A.J. Carlyle, *A History of Medieval Political Theory in the West*, 6 vols. (Edinburgh: Blackwood, 1903-36) 5:94, with Charles H. McIlwain, *The Growth of Political Thought in the West* (New York: Macmillan, 1932), p. 331, n. 1.

6. Aquinas relates human law to custom with considerable sophistication. ST 1–2. 97. 2 and 3. While Aquinas does not directly address the problem of radical regime change, his embryonic theory of popular consent provides a theoretical basis to justify modern democratizing revolutions. The principle of self-determination, of course, is difficult to apply where the political community with a right to determine its regime is precisely the subject matter in dispute. See my article, "Aquinas on Political Obedience and Disobedience," *Thought* 56 (March 1981):81–82, 86–88.

7. Aquinas' views on the relation of temporal to spiritual authority is beyond the scope of our concerns. Suffice it to say that Aquinas assigns supreme temporal and spiritual power to the Pope, although the Pope wields no immediate temporal power outside the Papal States. There is one *respublica Christiana* with one head. See CS 2. 44 and 4. 37.

In this context of a sacral society, Aquinas favors only limited political tolerance of nonbelievers and very extensive political intolerance of heretical Christians. Jews and pagans, who have never professed the Catholic faith, should in no way be compelled to embrace that faith. But they should be compelled to cease blasphemies, persuasions, and persecution (ST 2-2. 10. 8). Nonbelievers should not institute rule over Christians, although they may continue to rule over Christians where their rule preceded Christianization of the realm, unless the Church were to abrogate that rule (ST 2-2. 10. 10). The rites of Jews should be tolerated but not those of other nonbelievers (ST 2-2. 10. 11). On the other hand, heretics should be "physically forced" to fulfill what they promised and to profess what they once accepted (ST 2-2. 10. 8). Heretics merit death, and the Church should let secular authorities administer that fate to persistent heretics (ST 2-2. 11. 3). Aquinas was thus hardly a model of ecumenism. Two points, however, may be noted: (1) Aquinas erroneously assumed as a matter of culturally conditioned political theory that the ends of society and those of government in religious matters coincided; (2) Aquinas erroneously assumed as a matter of fact the bad faith of all heretics and the vincible ignorance of many nonbelievers.

8. The principle of hereditary legitimacy itself is not self-defining, as the pages of medieval history eloquently testify. Nor are the constitutional customs limiting royal power in Aquinas' theory very clear, again as the pages of medieval (and modern) history testify. Nor is it always easy to distinguish when medieval kings acted for their private good rather than the common good, especially when there was no separation of the king's privy purse from the public exchequer.

9. Cf. Robert M. Hutchins, *St. Thomas and the World State* (Milwaukee: Marquette University Press, 1949).

10. Capitalization, punctuation, and spelling of the text of Hobbes have been modernized.

11. Scholars basically divide into three camps on the role of morality in Hobbes' system. One camp denies any truly moral dimension to Hobbes' universe. See, e.g., George Sabine, *History of Political Theory* (New York: Holt, Rinehold, and Winston, 1937), p. 469, and Bertrand de Jouvenal, *Sovereignty*, trans. J. F. Huntington (Chicago: University of Chicago Press, 1957), p. 242. A second camp claims that moral obligation in Hobbes is genuine, but that moral obligation exists only in the commonwealth. See, e.g., Sterling Lamprecht's introduction to Hobbes' *De Cive* (New York: Appleton-Century-Crofts, 1949), pp. xxiii–iv, and Michael Oakeshott's introduction to Hobbes' *Leviathan* (Oxford: Blackwell, 1946). The third camp argues that Hobbes morally obliges humans in both the state of nature and the commonwealth, although the extent of moral obligation differs in the two conditions. See, e.g., A. E. Taylor, "The Ethical Doctrines of Hobbes," *Philosophy*, 13 (1938): 406–24, and Howard Warrender, *The Political Philosophy of Hobbes: His Theory of Obligation* (Oxford: Clarendon Press, 1957). While a case can be made for the first two positions, the third position seems the most plausible, provided that one distinguishes the morality of intention in the state of nature from the morality of action in the commonwealth.

12. This interpretation of Locke's law of nature is at least consistent with his interpretation of Christianity. Locke argues in his *Reasonableness of Christianity* that Jesus' ethical teachings are consistent with reason. Locke was a Socinian, i.e., what moderns would call a Unitarian. Locke's rejection of orthodox Christianity would suggest that he accepted the providence of God as Creator on the basis of some sort of natural faith.

13. Locke relates the natural right to property to the natural right to life. Every human has a right to his or her own person and labor (TG 5. 27). This natural right extends to whatever a human produces for self and family (TG 5. 31) and to land cultivated for that purpose (TG 5. 32). But humans have no natural right to accumulate wealth; the right to accumulate wealth derives rather from mutual consent to recognize money as the unit of exchange (TG 5. 47).

14. Marx also maintained that humans originally lived in a state of primitive communism somewhat akin to the state of innocence of Adam and Eve before the Fall, and privatization of property was Marx's equivalent of original sin in Christian theology.

15. There are to be two stages of the end-time in Marx's theory. The first stage will be socialism. In that stage, workers will control the means of production, but a dictatorship by the party elite will be necessary to reconstruct society and social consciousness. The second and definitive stage will be communism. Workers will then not only control the means of production but also have a correct social consciousness.

16. There is a fundamental paradox, if not contradiction, between Marx's call for workers to unite to take over the means of production and his economic determinism. If economics determines human behavior, how can workers do anything else but what they do? On the one hand, Marx seems to think workers capable of deciding to organize to overthrow the capitalist system. On the other hand, Marx insists that economics determines all human behavior. Critics have long noted this seeming contradiction.

17. Thomas P. Neill, *Makers of the Modern Mind*, 2d ed. (Milwaukee: Bruce, 1958), p. 299.

18. I summarize in this and the next six paragraphs themes excellently elaborated by Jacques Maritain, "The Person and Society," *The Person and the Common Good* (Notre Dame, IN: University of Notre Dame Press, 1966), pp. 47–89.

19. In fact, according to Marx, production under communism will become so abundant that workers will be able to satisfy their desires as well as their needs.

3

Political Obligation and Justice

Political Obligation and Institutional Justice

If humans naturally incline to organize themselves into a community, a body politic, and if political authority is the supreme organ of the body politic, it follows that humans should obey the just laws of just regimes (CS 2. 44. 2. 2; ST 2-2. 104. 6).[1] According to Aquinas, human laws are just and binding in conscience if they satisfy three conditions (ST 1-2. 96. 4). First, with regard to their finality, human laws should be directed toward the common good. Second, with regard to their authority, human laws should proceed within the limits of the lawmaker's constitutional power. Third, with regard to their form, human laws should lay burdens on citizens with a proportional equality.

These three characteristics that Aquinas ascribes to specific laws may serve as a point of departure for considering the justice of political institutions generally. The primary goal of organized society is to promote human *self*-development, and so freedom should be a characteristic of just regimes. Lawmakers derive their authority to govern from the fundamental constitution of society, and so popular consent should be a characteristic of just regimes. And because organized society is ordered to the good of the citizens who compose it, citizens should be treated as equals under a rule of law.

"Popular consent" is a relative term. Widespread and meaningful participation in a democratic political process would obviously indicate a high degree of popular consent to the exercise of political authority. But citizens may support a nondemocratic regime too. Tribes, for example, may defer willingly to the authority of tribal elders. The sticking point comes when societies are in a state of transition from traditional to modern organization.[2] In such conditions, popular consent to the traditional regime is likely to become so eroded that at some point it will be insufficient to justify obedience to that regime. Just exactly when that point is reached will be evident after the regime is overthrown, but it may be evident enough well before that occurs.

Freedom represents an ideal which political institutions realize in different ways and to different degrees. As human persons, citizens should be free to act according to their own will as much as possible, and political institutions should coerce citizens to act against their will only as much as necessary for the well-being of the community. But there will be different interpretations about what is necessary for the well-being of the community. A democratic society resolves those differences by the ballot box; a traditional society resolves them by decisions of the governing elite. A democratic society is likely to satisfy the criterion of freedom to a substantial degree. (This is not necessarily the case, however; the United States, for example, permitted slavery for nearly the first hundred years of its national existence.) A traditional society may also minimally satisfy the criterion of freedom if the governing elite acts moderately. In a state of transition from traditional to modern society, however, the governing elite is not likely so to act.

Equality is perhaps the most relative of the criteria according to which the justice of political institutions is to be judged. At a minimum, equality should mean that citizens are treated equally under a rule of law, and that political, social, and economic inequalities need to be justified in terms of the well-being of the community. One citizen should not receive preferential treatment over another in the administration of laws because of his or her personal characteristics (race, religion, sex, blood relationship, friendship). And where political, social, or economic inequalities do exist, they should be justified by objective and rational factors related to the good of the entire community. The very notion of organized society implies some inequality. Organization means that some persons will be in charge of some things, and those persons, by reason of their position, will have a higher status than other persons. This is true of the Soviet Union as well as a monarchy like the United Kingdom, although there may be more social and economic inequality in Great Britain than in the Soviet Union. The British generally accept the privileged position of Queen, Lords, and social elites and justify the privileges by reason of the recipients' service to the community. One might add incidentally that symbols and customs themselves are perfectly rational considerations when judging the utility of inequalities for the common good.

Freedom and especially equality are general concepts whose implications for different situations citizens can recognize only by rational reflection and argument. We take for granted today that persons of all races, all religions, and both sexes should be free and treated equally. Recognition of the universal applicability of the principles of freedom and equality is, however, of relatively recent vintage. Athenian freedom and equality extended only to male citizens and accepted the institution of slavery. Medieval society excluded women from positions of political, social, and economic power and accepted the institution of serfdom. The United States did not extend the suffrage to women until 1920 and did not abolish slavery until 1865. In fact, the United States did not prohibit racial segregation in public education until 1954 and did not dismantle the bulk of segregated Southern school systems until the mid and late 1960s. Today, the Western world has largely eliminated legal discrimination against non-Caucasians, Jews, and women.

But obviously much remains to be done in the cause of racial and sex discrimination at social and economic levels. Whether or not particular policies to remedy such discriminations, like affirmative-action programs, are properly directed against the evils to be corrected or otherwise desirable, are matters for legitimate public debate. On the one hand, it can be argued that affirmative-action programs penalize innocent individuals in order to benefit victims of past discrimination. On the other, it can be argued that society itself was responsible for the past discrimination, and so that no member of the classes which were the beneficiaries of past discrimination can plead total innocence.

Rawls

The arguments advanced here that individual citizens and groups should enjoy as much freedom to act according to their will as is compatible with the good of the community, and that political, social, and economic inequalities must be justified in terms of the good of the community, bear some resemblance to John Rawls's theory of justice. There are, however, significant differences between the two lines of argument.[3] Rawls hypothesizes a situation in which free and rational individuals are in an "original position" of absolute equality corresponding to the state of nature in the theories of Hobbes and Locke, and he stipulates that no one in this situation knows his or her future status in society, his or her share of natural assets, or even his or her own conception of the good. Rawls argues that individuals in this position would rationally agree to two principles of institutional justice as integral to agreement to enter an organized society: (1) each individual would have an equal right to the most extensive freedom compatible with a similar freedom for other individuals; (2) social and economic inequalities would be permitted only if they can be expected to benefit everyone and would be attached to positions open to all.

Rawls's "game theory" model is explicitly directed against the utilitarian ethical theory current in many American universities, and it is implicitly directed against laissez-faire economic theories dominant in American business. Within that context, Rawls plausibly argues that the equality postulated for a state of nature by traditional social contract theorists has broader implications than those theorists recognized. At least Rawls's argument should induce his utilitarian adversaries to admit inequalities in the state of nature if they wish to avoid acceptance of the principles of justice that Rawls deduces from the "original position."

From the Thomistic perspective, however, we may question Rawls's starting point and his restricted notion of human reason. Rawls's premises are those of the Enlightenment. Individuals are assumed to be uninterested by nature in the company or well-being of others, that is, individuals are assumed to have no natural inclination to live in a community. Rawls's concept of practical reason is one of calculating how best to achieve subjectively desired benefits without any ability to relate those subjective benefits to an objective order of goods perfective of human nature. If one believes, however, that humans in fact naturally do incline to cooperate with their fellows, and that human reason is capable of understanding from that fact that humans should do so, Rawls's project is basically unnecessary. Rawls himself accepts as simply given intuitive human convictions about the primacy of justice in sociopolitical relations. Thomists would ground those convictions in human nature and recognize them as more than mere assumptions.

The further question is whether or not Rawls's theory of justice is sufficient. When Rawls deals with the nitty-gritty of what inequalities his theory would justify, he is unable or unwilling to commit himself. He admits, for example, that a case could be made for qualifications on voting weighted to reflect intellectual ability (à la John Stuart Mill). And Rawls does not say what degree of social or economic inequality he would justify as ultimately beneficial to the worst off in society. In short, we need to use practical reason to determine the qualitative as well as the quantitative consequences of political institutions. This reinstates *sub rosa* the role of reason that Rawls excluded from the "original position," wherein the principles of justice were derived.

As a result of Rawls's individualistic premises and the restricted role he gives to reason, his formulations of the principles of justice differ from those we have suggested. According to Rawls, each individual should have an equal right to the most extensive freedom compatible with similar freedom for other individuals. According to Thomistic theory, humans have natural rights to as much freedom as is compatible with the good of the community, which good includes the rights of individual citizens but is not restricted to protection of those rights. We have argued that the freedom of individual citizens and groups is an important good, perhaps the most important good, of the community. But the good of the community goes beyond maximizing the freedom of individuals to do whatever they wish to do; the good of the community includes the whole material and spiritual environment in which individual citizens exercise their freedom.

According to Rawls, social and economic inequalities should be permitted if and only if the inequalities can be expected to benefit everyone minimally. According to Thomistic theory, humans have a natural right to receive social and economic benefits proportionately equal to their contribution to the good of the community. We shall explore this subject more fully in Chapter 5, "The Justice of Economic Institutions and Distributions." Here we note only the chief difference between Rawls's formulation and the Thomistic formulation. Rawls finds justice satisfied as long as those worst off in society are minimally better off with the inequalities than they would be without the inequalities. Thomistic theory would not find justice satisfied simply because the worst off gain something from inequalities; the worst off might deserve a bigger slice of the pie than the minimum with which they can be said to be better off. Rawls uses a quantitative formula to answer a qualitative question. (One might also quarrel with Rawls's stipu-

lation that justice would only be satisfied if social and economic
inequalities are attached to positions open to all. Taken literally,
that would presumably make even a constitutional monarchy like
that of the British unjust.)

Areas of Political Justice

One area of political justice involves crime and the punishment of
crime. Any organized society needs to regulate the behavior of
citizens in order to protect and promote the well-being of the
community and its members. When individuals violate those rules,
they act wrongly against the good of the community and its
members and deserve to be punished. Justice requires that the laws
defining criminal acts be clearly articulated and related to the
common good, that the punishment assigned to criminal acts be
proportional to the harm done to the community, that the process
for determining guilt or innocence be rational and fair, and that
the process be followed in practice. Miscarriages of justice will
inevitably occur in any criminal justice system, but the system
itself will be just if the substance of criminal laws and the process
of administering them are just.

Imprisonment or death can be justified only as punishment for
wrongs done to the community and its members. The argument
that capital or any other punishment is necessary to deter crime is
irrelevant from the viewpoint of justice. Indeed, disproportionate
punishment of wrongdoers or any punishment of innocent citizens
in order to deter crime would be unjust. Undoubtedly, punishment
of wrongdoers will have a deterrent effect, and the more severe the
punishment the more crime might be deterred. But the punishment will only be just if the individual deserves to be punished,
and if the punishment is proportionate to the wrong done to the
community. Similarly, prisons may help to rehabilitate the criminal, although that does not presently seem to be the case, at least in
this country, but the justice of imprisoning the criminal at all
depends on the justice of punishing wrongdoing.

Political justice involves regulating the activities of citizens in
other ways than that of designating activities as crimes. Market
societies will need to regulate commercial activities, settle property
questions, and adjudicate disputes between citizens about the violation of rights. Contract law has developed a very sophisticated set
of rules: contracting parties must be of age, competent, and un-
coerced; promises by the contracting parties must involve some

cost or consideration on the part of each; contracts with respect to the commission of criminal or immoral acts will not be enforced; fraud will void contracts; contracts will be interpreted liberally against parties who virtually dictate terms of the contracts. Similarly, property law has developed an elaborate set of rules to regulate the title and transfer of property by sale and inheritance.

Tort law, law concerned with the violation of one citizen's rights by another, resembles criminal law insofar as it defines what is right or wrong civic behavior; it differs from criminal law in that it defines what is right or wrong behavior of citizens toward one another rather than what is right or wrong behavior of citizens toward the community. Physical damage to the person or property of another, trespass, and libel are examples of torts. Tort law will define what constitutes a wrong by one citizen or group against another and what compensation is due to the other when a wrong has been committed. (The same wrong, of course, may be both a crime and a tort.) Tort law differs from criminal law in one respect other than the public or private character of the injured party. In criminal law, the wrongdoer must deliberately intend to do wrong; in tort law, some agents will be liable for damages even though they did not intend to do wrong (e.g., unwitting trespass), and other agents will be liable for damages simply because they should have known the consequences of their acts (e.g., negligence).

The justice of economic institutions, like the forementioned types of civil law, is concerned in part with defining the rights and duties of individuals and groups in organized society (e.g., corporate law, labor law). But it is also concerned with government activity relating to individuals and groups (e.g., welfare law, social security.)

Moral Obligation and Unjust Laws

The above discussion related political obligation to the overall justice of regimes. Here we shall discuss the moral problems posed by specific laws or policies of regimes substantially satisfying justice. Even though regimes may substantially satisfy the three suggested criteria of institutional justice, particular laws, policies, or their application may be unjust. The question then arises whether or not citizens are morally obliged to obey such laws.

Let us return to Aquinas's criteria for just laws. The first criterion requires that just laws be directed to the common good of the body politic, not the private good of the lawmaker (e.g., the lawmaker's cupidity or vainglory). The second criterion requires that human laws not exceed the constitutional power of the law-maker. The third criterion requires that human laws not lay dis-proportionate burdens on different citizens and classes of citizens even if the burdens are ordered to the common good. Violation of any one of these criteria will render a law unjust. Such laws do not bind in conscience, "except perhaps to avoid scandal or distur-bance, for which cause a man should yield even his right" (ST 1-2. 96. 4).[4] These laws often bring injury (i.e., unjust harm) to individ-ual citizens, and citizens so affected may seek to avoid oppression and violence, provided they avoid scandal and the imposition of more grievous harms on the community or other citizens (ST 1-2. 96. 4. ad 3).

Aquinas refers to situations in which civil laws unjustly inflict harm on individual citizens. The laws at issue do not compel the adversely affected citizens to commit any injustice, only to suffer it. The narrow question is whether or not citizens so affected are morally obliged to obey the laws and thereby suffer the loss. The answer Aquinas gives is that the affected citizens have no moral obligation to obey unjust laws inflicting harm on them simply because the lawmaker has enacted the laws. Just laws oblige in conscience only because each citizen is part of the community and as part is subordinate to the proper good of the whole community. Just laws inflict losses on parts of the community in order to protect and promote the good of the whole community. But laws which are not designed for the good of the community, or which exceed the authority given by the community to the lawmaker, or which impose disproportionate burdens on some members of the community cannot make any moral claim on individual citizens to sacrifice themselves for the good of the community.

Aquinas notes, however, that an unjust law inflicting harm on an individual citizen may be morally obligatory for reasons extrin-sic to the justice of the law. He summarily cites avoidance of scandal as one extrinsic reason why an adversely affected citizen may be morally obliged to obey an unjust law. The argument, not articulated by Aquinas, would run somewhat as follows. Because the unjust law has the appearance of law, disobedience of the law might lead other citizens to disobey just laws, undermine the rule of law, and so harm the community. The other example cited summarily as an extrinsic reason why an adversely affected citizen may be morally obliged to obey an unjust law is the avoidance of civil disturbance. Disobedience of an unjust law might lead to an armed conflict between a ruler and his subjects and so harm the community. In such cases, the individual "should yield even his right." It is thus not enough for individual citizens to appeal to their "rights" in the matter of political obedience; the good of the community takes precedence over the rights of individual citizens on whom unjust laws inflict harm.

While laws may be unjust because they inflict harm unjustly on citizens, they may also be unjust because they compel citizens to act contrary to the divine law, and "laws of this kind must nowise be observed" (ST 1-2. 96. 4). By divine law, Aquinas understands both divine positive law and divine natural law (ST 2-2. 57. 2. ad 3). Divine positive law embodies those commands and prohibitions made by special divine decree; the things commanded are good because God prescribes them, and the things prohibited are bad because God forbids them. Divine natural law embodies those commands and prohibitions implicit in the created order and so accessible to human reason without any special divine revelation; the things commanded are prescribed because God ordained the natural order which constitutes them good, and the things prohibited are forbidden because God ordained the order which constitutes them bad.

Civil laws would run counter to divine positive law and so be unjust if, for example, the laws commanded a divinely disapproved form of worship or prohibited a divinely approved form of worship. Civil laws would run counter to divine natural law and so be unjust if, for example, the civil laws commanded citizens to worship God in ways contrary to their conscience, or if the civil laws prohibited citizens from worshipping God according to dictates of their conscience in ways compatible with public order. Thus, a civil law commanding *any* form of worship or forbidding *any* form of worship compatible with public order would run counter to either divine positive or divine natural law or both and so be unjust. For our present purposes, we shall consider exclusively the question of moral obligation to obey laws that are unjust because they are contrary to divine natural law.

First, we should note that the context is one of conflict between what civil law commands or prohibits and what the natural law prohibits or commands. That is, the context is one of conflict between civic legal duty and natural moral duty, not between civic duty and natural right. Civil laws may unjustly inflict harm on citizens, and we have previously considered that possibility and its implications for moral obligation. In such cases, adversely affected citizens are threatened with unjust harm, but they are not commanded to do anything that the natural law prohibits nor are they prohibited from doing anything that the natural law commands. The conflict there is between what the civil law commands or prohibits and what the individual has a natural right not to do or to do. The conflict here is between what the civil law commands or prohibits and what the individual has a natural duty not to do or to do. Unjust denials of natural rights compel no immoral action or inaction on the part of adversely affected citizens.

Second, we should note that the conflict between requirements of civil law and natural law typically involves a situation in which civil law commands an action contrary to natural law. Let us take an example from warfare. Civil authorities may command citizens to fight in an unjust war, while the natural law prohibits citizens from doing so. In that case, citizens would not only be under no moral obligation to wage unjust war, they would have a positive moral obligation not to wage such war. This principle would apply even to the moral obligation of a citizen to obey an objectively just law commanding an action that a citizen erroneously but sincerely believes prohibited by divine natural or divine positive law. A pacifist, for example, sincerely believes that all wars are contrary to divine natural or divine positive law. Assuming that some wars can be justified, and that a particular war is justified and fought by just means, civic authorities would justly require citizens to wage such war. The pacifist, however, not only is under no moral obligation to do so but even has a moral obligation not to do so.[5]

Conflicts between civic and moral duty raise a host of legal and political problems, especially in cases of those who claim conscientious objection to particular wars.[6] How do civic authorities distinguish sincere from insincere claims? How do civic authorities distinguish morally motivated from politically motivated objection to particular wars? How can civic authorities accommodate selective conscientious objectors to particular wars and still allocate the burdens of war fairly to all citizens? How do civic authorities discourage selective obedience to law if they make an exception for selective conscientious objectors to particular wars? These are serious but not insuperable problems. Civilian alternatives to military service are one obvious practical solution, and that is the recourse the United States has followed in one form or another from colonial days, in the case of religious pacifists, and, in recent years, in the case of secular pacifists as well.[7] Unfortunately, that solution has not been adopted to accommodate selective conscientious objectors to particular wars.[8]

Citizens may object not only to the performance of acts specifically deemed to be immoral but also to associated acts. Pacifists, for example, may morally object to alternative civilian service or even to registration for the draft as well as to all forms of military service.[9] Such objections would be based on a perception that alternative civilian service or registration for the draft constitutes morally impermissible cooperation with the waging of war. The degree to which individuals may cooperate with activities deemed immoral without incurring moral fault is a difficult matter to determine both objectively and subjectively. Objectively, reason needs to determine when associated acts are so closely linked to acts deemed immoral that performance of the former is equivalent to performance of the latter. Subjectively, even within the same set of religious beliefs or philosophical principles, individuals may reach different conclusions about the linkage. In any case, if an individual reaches a firm conclusion that he or she is not morally permitted to perform a particular act because of its linkage to another act, that individual would be morally obliged to follow his or her conscience. (Of course, the further individuals press their conscientious objections from the specific acts deemed immoral to associated acts, the more doubt may be cast on the sincerity of representations of the state of their conscience.) Civic authorities, however, may be unable to accommodate all individual perceptions that associated acts are morally prohibited without sacrificing important public interests.

The situation where civil law prohibits an action which the natural law categorically commands is atypical. One can imagine a case where civil law would prohibit an action, for example, trespass on public property, and a citizen would perceive a categorical moral duty to perform the action, for example, trespass on public property to save a human life. Such a situation is unlikely because civil law itself usually makes room for the exceptional case. In other words, a rule of reason usually operates in the application of civil law. More fundamentally, civil law rarely prohibits an action which an individual's conscience categorically commands. The conflict between civil laws prohibiting actions and individuals' perceptions of moral duties is typically one in which the individual perceives only a generalized duty to act. Individuals may perceive a duty to preach their religious message, for example, but that perception involves no consciousness of a duty to preach the message in specific ways; the time, place, and means of implementing the duty to preach a religious message are left to individual determination, and if specific modes of implementation are legally prohibited, other modes remain available. It will be a rare case where use of a specific means is perceived as a moral duty.

Occasionally, individuals may perceive relatively specific religious or moral duties. Mormons in the second half of the nineteenth century, for example, claimed that they had a divine positive mandate to practice polygamy. Had the Volstead Act made no provision exempting sacramental wine from the prohibition of alcoholic beverages, Roman Catholic priests might have similarly claimed a divine positive mandate to manufacture and transport wine in order to celebrate mass. Even in those cases, however, the consciences of those affected by the civil laws would not have categorically commanded disobedience of the laws. The Mormon mandate explicitly specified "circumstances permitting."[10] And Roman Catholic moral theology teaches that mass can only be celebrated if the requisite conditions are satisfied (by an ordained priest, with bread and wine). The absence of women, in the one case, or the absence of wine, in the other, would absolve Mormon males and Roman Catholic priests from any moral obligation to practice polygamy or celebrate mass, respectively. The moral constraint is serious, but it is not a conflict between civil legal duty and strict moral obligation.

Civil Disobedience

In addition to moral objections to specific laws, there is the relatively recent phenomenon of civil disobedience. Practitioners of civil disobedience claim that they are morally justified, perhaps even morally obliged, to disobey federal or local laws in order to change public policies affecting minorities, the poor, or the waging of war. There are two central elements in civil disobedience: civility and disobedience. We shall consider each element in turn.

The term "civility" poses particularly difficult problems of philosophical analysis and definition, about which practitioners and commentators disagree. Should the act of disobedience always be performed openly? Should the act of disobedience be planned or may the act be spontaneous? How concrete should the targeted public policy be? How much should civil disobedients do to exhaust legal remedies before resorting to acts of disobedience? May civil disobedients ever use violent means to achieve a change in public policy? To what extent should civil disobedients respect the person and property of others in the course of acts of disobedience? How closely linked should the law disobeyed be to the targeted public policy? Should civil disobedients voluntarily submit to the legal consequences of their acts? For our purposes, it is unnecessary to resolve these questions, except to note that the answers to the questions, especially with regard to the means employed, will increase or decrease the effects of disobedience on the public.

The term "disobedience" has a number of meanings or applications with respect to the practice of civil disobedience. One meaning or application might be disobedience of a federal or state statute that disobedients claim violates the U.S. Constitution. Most challenges to local laws by civil rights partisans in the early sixties, for example, claimed that the statutes involved mandated racial segregation in violation of the Fourteenth Amendment of the federal Constitution, or that the same amendment prohibited enforcement of private racial segregation by public authorities. In Thomistic terms, civil disobedients of the early sixties morally justified their disobedience on the ground that lawmakers mandating segregation or courts enforcing it exceeded their authority, and that the statutes or court orders were accordingly unjust. Subsequent Supreme Court decisions vindicated that position.[11]

A second meaning or application of the term "disobedience" in the context of civil disobedience might be disobedience of constitutionally valid laws that civil disobedients regard as substantially unjust. They might regard statutes as unjust because the statutes lay disproportionate burdens on different groups of citizens. Morally motivated disobedience of statutes mandating racial segregation in public facilities at a time when the Supreme Court interpreted the Fourteenth Amendment to allow such statutes as long as equal facilities were provided for non-Caucasians would be an example of this meaning or application. Or civil disobedients might regard statutes as unjust because the statutes command actions prohibited by the natural law. Morally motivated disobedience of statutes mandating military service for conscientious objectors to the Vietnamese war would be an example of this meaning or application.

A third meaning or application of the term "disobedience" in the context of civil disobedience might be disobedience of constitutionally valid laws protecting government personnel and property where the personnel or property is devoted to uses civil disobedients regard as immoral. Such was the case when opponents of the Vietnamese war attempted to block troop trains by lying on railroad tracks or when current opponents of nuclear weapons attempt to disrupt the weapons' production or deployment by trespass actions or token attacks on the weapons.

A fourth meaning or application of the term "disobedience" in the context of civil disobedience might be disobedience of admittedly constitutional and just laws in order to protest another law or policy that is regarded as unjust or immoral. Lying down on roadways during rush hours to protest the inadequacies of welfare programs would be an example of this type of civil disobedience. Civil disobedients would morally justify their disobedience of admittedly constitutional and just laws on the ground that ordinary means of political protest and activity are inadequate to alter the public policies to which they object.

The first two types of disobedience (disobedience of unconstitutional statutes; disobedience of statutes laying disproportionate burdens on different groups of citizens or commanding actions prohibited, or sincerely thought to be prohibited, by the natural law) can be justified on the Thomistic principles developed in the previous section of this chapter. But the last two claims of moral justification for (1) disobedience of admittedly constitutional statutes protecting government personnel and property on the ground that the personnel or property is linked to public policies deemed immoral or (2) disobedience of admittedly constitutional and just statutes on the ground that such disobedience is necessary to alter public policies deemed morally inadequate are another matter. These claims are quasi-revolutionary in nature and attack the institutional justice of the regime itself. The claims may well be valid, but their nature needs to be recognized for what it is.

The first two types of civil disobedients basically argue only that the government, as a matter of justice, should leave them alone, either because the government seeks unjustly to deprive them of their rights or property, or because the government seeks unjustly to coerce them to act in ways deemed immoral. The last two types of civil disobedients argue that the government, as a matter of justice, should alter its policies affecting the general public. In the former cases, disobedients claim a moral right or duty not to do something; in the latter cases, disobedients claim a moral right or duty to act against the government or the public in order to remedy the perceived injustice of public policies. Indeed, since the latter disobedients feel themselves unable either to alter public policies deemed immoral by rational argument or to persuade the mass of their fellow citizens by ordinary political processes, at least in the short run, they may perceive various acts of disobedience (nonviolent to violent) as relatively specific moral obligations. But whatever the degree of moral obligation the second two types of disobedients perceive, and however justified their perceptions, their acts constitute threats to the legal system and the public interest in general observance of law. Such acts—however civil—directly challenge not only the legal system but also the political system itself.

The National Commission on the Causes and Prevention of Violence was apparently referring to such acts when it defined the purpose of civil disobedience as to "influence community action by harassing or intimidating the members of the community into making concessions to a particular point of view."[12] The Commission admitted that the available evidence was insufficient to demonstrate that acts of civil disobedience of a more limited kind inevitably lead to an increased disrespect for law or propensity toward crime on the part of the disobedients. But the Commission did find that acts of civil disobedience, even of a limited kind, have a seriously adverse effect on the law-abiding behavior of other citizens and, therefore, on the system of law upon which organized society depends.[13] Citing India as an example, the Commission argued that civil disobedience tends to undermine the operation of democratic processes even when the disobedients do not intend that result.

Two purely practical points are worth noting about civil disobedience which seeks to alter public policies. The first is that civil disobedience to alter domestic policies, in proportion to its incidence, indicates that political authorities are not satisfying the aspirations of racial minorities or the poor. Although political authorities may have little hope of compromising differences with civil disobedients opposed to military or foreign policies, they may be able partially to satisfy the demands of civil disobedients opposed to domestic policies. The second point is that civil disobedience, in proportion to the degree it inconveniences the public, may be counterproductive. Harassing activity is not well-suited to persuading the public of the justice of the disobedients' cause, and such activity runs the risk of causing a public backlash against the disobedients' cause.

Notes

1. Aquinas deals with the right to revolution against unjust regimes only in the context of resistance to tyrants. On his thought on the subject and the stages of its development, see my article, "Aquinas on Political Obedience and Disobedience," pp. 82–85.

2. Talcott Parsons distinguishes traditional from modern societies by five patterns of action orientation: ascription vs. achievement, particularism vs. universalism, affectivity vs. affective neutrality, diffusiveness vs. specificity, and self-orientation vs. collective orientation. Talcott Parsons and Edward A. Shils, eds., *Toward a General Theory of Action* (New York: Harper and Row, 1951). Traditional societies accord status to individuals in society on the basis of age, sex, lineage, or other qualities unrelated to performance. Modern societies, on the other hand, accord status on the basis of achievement. Traditional societies accord favored treatment to members of the same family, clan, tribe, religion, or ethnic group. Modern societies, on the other hand, apply rules to all similarly situated individuals without discrimination. Traditional societies judge and reward individuals on the basis of personal likes and dislikes. Modern societies, on the other hand, judge and reward individuals in a distinterested way according to objective criteria. In traditional societies, individuals interact at general levels of friendship, family, and kinship. In modern society, on the other hand, contracts specify how individuals should interact. And members of traditional societies prefer individual and family goals to broader collective goals, while members of modern societies will sacrifice personal goals to community goals. These patterns of action orientation are not entirely free of ambiguities, nor are societies entirely traditional or modern. But the patterns do serve in a general way to distinguish undeveloped from developed societies.

3. John Rawls, *A Theory of Justice* (Cambridge, Mass.: Belknap, 1971), pp. 60–75.

4. Cf. CS 2. 44. 2. 2.

5. Moral obligation not to perform an act should be distinguished from moral obligation not to submit to an act performed by others. Jehovah's Witnesses, for example, perceive a moral obligation not to obtain blood transfusions, but they may perceive no moral obligation to resist court-ordered doctors administering blood transfusions. See *Application of President and Directors of Georgetown College, Inc.*, 331 F.2d 1010 (1964) and *U.S. v. George*, 239 F.Supp. 752 (1965). Acts mandated by public authorities but performed by others than those with moral objections to the acts need not be perceived by the affected individuals as involving any action on their part contrary to their conscience. Nonresistance and submission to the power of others might involve only passive cooperation with acts deemed immoral.

6. See, generally, Richard J. Regan, S.J., *Private Conscience and Public Law: The American Experience* (New York: Fordham University Press, 1972), pp. 33–49.

7. Ibid., pp. 21–33.

8. *Gillette v. U.S.*, 401 U.S. 437, 91 S.Ct. 828, 28 L.Ed.2d 168 (1971). The Court there declined to follow the "alternative means" approach it explicitly recognized in the case of Amish parents objecting to compulsory public education of their children beyond the eighth grade. *Wisconsin v. Yoder*, 406 U.S. 205, 92 S.Ct. 1526, 32 L.Ed.2d 15 (1972).

9. For a case involving a pacifist who objected to alternate civilian service, see *U.S. v. Boardman*, 419 F.2d 110 (1969).

10. See *Reynolds v. U.S.*, 98 U.S. 145, 25 L.Ed. 244 (1879).

11. *Peterson v. Greenville*, 373 U.S. 244, 83 S.Ct. 1119, 10 L.Ed.2d 323 (1963); *Lombard v. Louisiana*, 373 U.S. 267, 83 S.Ct. 1122, 10 L.Ed.2d 338 (1963); cf. *Bell v. Maryland*, 378 U.S. 226, 84 S.Ct. 1814, 12 L.Ed.2d 822 (1964).

12. *To Establish Justice, to Insure Domestic Tranquility: The Final Report of the National Commission on the Causes and Prevention of Violence* (New York: Bantam, 1970), p. 86.

13. Ibid. All members of the Commission agreed to this finding, but they split sharply on the moral justification and ultimate consequences of civil disobedience.

4

Moral Good and Civil Good

Modern minds typically think of moral virtue as a purely individual and private matter. Aristotle and Aquinas, on the other hand, thought of moral virtue as quintessentially social and of public significance. The purpose of this chapter is to consider the relation of the moral virtue of individual citizens to the common good, the essential role of consensus on moral good in shaping public policies, and the function of legal coercion to enforce public morals. In this context we shall examine the specific problem of abortion legislation and public morals.

Ancients like Aristotle and medievals like Aquinas situated humans in a morally ordered universe. For them, there exists an objective order of goods perfective of human nature, and this order of goods is normative for individuals. Among the goods perfective of human nature is fellowship, and so humans naturally incline to live in an organized society. Because human persons perfect themselves not only by physical activity but also and especially by intellectual activity and the moral character of their free acts, the good of organized society includes citizens' intellectual and moral well-being as well as their physical well-being. Like individuals, society should be rightly ordered to moral good. Negatively, as we noted in the last chapter, this means that the political authority of a rightly ordered society should not command or induce citizens to do what is morally bad or prohibit them from doing what is morally obligatory. But moral good has a much more important role to play in organized society than merely to provide a negative norm to distinguish morally right from morally wrong exercise of political authority. Political authority is entrusted with the specific function of developing the good of the whole community, and the good of the community includes its moral well-being. The "good life" is the morally good life, and political authority is responsible for the public climate in which citizens develop themselves morally. Aristotle expressed this view of the linkage of political goals to the moral development of citizens thus: "Lawgivers make citizens good by inculcating [good] habits in them, and this is the aim of every lawgiver; if he does not succeed in doing that, his legislation is a failure" (NE 1103^{b}3–5).[1]

Classical liberal theorists like Hobbes and Locke, on the other hand, trace the organization of human society and political authority to the decision of egocentric individuals to give up all or part of their freedom in order to guarantee their peace and security. In this view, organized society is a mutual nonaggression pact. Neither organized society nor the state has any strictly moral purpose; the only purpose of organized society and the state is to insure physical peace, not moral virtue. Liberal theory laudably aims to maximize individual freedom, but physical freedom divorced from all moral virtue, ironically, will not be sufficient even to secure long-term physical peace.

Americans for several generations have renounced the extreme liberal individualism of laissez-faire economics. Yet, despite the evidently moral dimensions of so many social and political crises in the late twentieth century, many Americans still cling to classical liberal individualism in the area of public morality. Today's extreme individualism in moral matters makes no more sense than yesterday's extreme individualism in economic matters. An organized society that will not abandon its economic destiny to the marketplace need not abandon its moral destiny to that marketplace either.

But if classical liberals and their contemporary heirs exaggerate the moral autonomy of individuals in the interest of freedom, conservative groups like the Moral Majority may exaggerate the role of political authority in the interest of the moral well-being of society. Political authority does not preempt the field of public morals. There are, in fact, many other units of organized society concerned about public morals besides government. The primary social unit concerned about morals is the family. The young receive their first and perhaps their most effective moral education from their parents; parents teach their children moral values by their example and their prudent exercise of discipline more than by their words. Second, schools share the task of educating the young morally, at least to the extent of the further development of social consciousness.[2] Third, churches and synagogues promote public morals insofar as they urge their members to observe moral codes. And voluntary associations promote public morals by their internal codes of ethics and by whatever good works they perform.

In a rightly ordered society, government should do for citizens only what individuals and lesser units of social organization do not or cannot do for themselves. While political authority in this scheme of things has the specific role of protecting and promoting the good of the whole, other units of society protect and promote the good of the whole when they pursue properly their own limited social goals. The main function of a rightly ordered political authority is to encourage subordinate units of society to pursue properly their own goals and so to promote common moral well-being. Thus, a rightly ordered society will seek to strengthen the roles of parents, teachers, preachers,[3] and those engaged in charitable works. In a rightly ordered society, legal enforcement of public morals should be the last resort of political authority, the last line of defense of the community's moral well-being.

With respect to the role of moral good as the overarching principle of political action, John Courtney Murray spoke of the "tradition of reason,"[4] Sir Ernest Barker of the "tradition of civility,"[5] and Walter Lippmann of the "public philosophy."[6] Key to the vitality of a public philosophy is a public consensus, that is, widespread civic agreement on basic rational principles of moral good to govern the activities of organized society and public policies. Libertarians tend to deny that such a consensus is necessary, desirable, or even possible in a free society. For them, the only absolute moral principle in a free society is commitment to freedom itself. Citizens, in a word, need agree only on the right to disagree.

At the outset, it should be noted that public consensus on basic moral principles in no way challenges principles of democracy. In fact, the contrary is true. The democratic process and its associated freedoms are themselves moral principles about the right ordering of the body politic that derive from the rational and free structure of human personality; democratic freedoms are not merely political means but indeed part of the ultimate goal of organized society itself. But the freedom of individuals and society to make decisions should not be confused with the substance of the decisions made.

The ultimate goal of politically organized society is to promote the human well-being of its members, and individual freedom is only part of that goal. To speak of human well-being is to speak at once of moral as well as physical well-being. This is precisely why a public consensus on moral good should inform the whole political process of decision making and the exercise of political freedoms.

Moreover, agreement on the democratic process and its associated freedoms by itself will scarcely insure enough unity of mind and heart to transform a collection of autonomous individuals into an organized society. Members of an organized society need to share common goals and aspirations, and the common goal and aspiration of freedom is not enough to give direction to organized society. Justice Frankfurter once spoke felicitously (albeit in the questionable context of compelling flag salutes from those with religious objections) of the "agencies of mind and spirit which serve to gather up the traditions of a people, transmit them from generation to generation, and thereby create that continuity of a treasured common life that constitutes a civilization."[7]

The public consensus concerns basic moral principles which should determine society's goals and the means society chooses to implement those goals. Pessimists are wont to speak in this regard of a descending curve of public morality. And in many areas, indeed, agreement on principles of public morality has visibly declined; the public today is obviously more indifferent than even a generation or two ago to the moral imperatives on such issues as extra- and premarital sexual behavior and abortion. Yet, in many other areas, agreement on principles of public morality has expanded; the public today is more concerned than previous generations with moral imperatives on such issues as war and peace, racial and sex discrimination, poverty, business practices, and the conduct of public officials. The resulting contemporary status of the public consensus on moral principles may be described summarily with respect to the proximate past as weaker in the area of personal and familial responsibilities and stronger in the area of social and political morality.

There is, therefore, occasion for both concern and hope over the present vitality and future prospects of public agreement on basic moral principles in America. The primary tasks for the American people and its leaders are to defend—more accurately, perhaps, to recover—the personal and familial territory of the consensus while continuing to advance its social and political frontiers. Yet, such tasks can be undertaken only if there is first general agreement on the centrality of moral good as common good and a general commitment to rational dialogue on problematics of the consensus. A tradition of reason requires a people, and especially its leaders in all fields, to accept the role of practical, moral reason in human society and to apply that reason to the problems of human society.

Religiously minded citizens believe that divine initiative is the sustaining force in the recognition and observance of even natural or secular morality. And cultural anthropologists confirm that religion operates to provide substantive moral codes and the motives to observe those codes. In America, too, our religious traditions are responsible in great part for the extent to which the public consensus on morals is an effective force. The ultimate key to the future vitality of the American public consensus on moral principles may well lie with the religious communities. Since religious communities and organized society itself share aspirations for moral good, religious communities should especially work to develop and elevate the public consensus on moral principles.

The Role of Coercion

We have insisted that the moral virtue of citizens is a goal, indeed one of the most important goals, of organized society and urged the importance of public agreement on basic moral principles as an operative force in American society. But to what extent should political authority use legal coercion to enforce standards of public morality?[8] This is an issue which deeply disturbs civil libertarians. Libertarians fear that even assent in principle to an objective order of moral goods with social dimensions will entail broadscale restriction of citizens' freedom. Libertarians do well to remind us that freedom itself is a moral goal of organized society.

Only a moral simplist would resort to legal coercion to secure every moral good. On the other hand, only an amoral simplist would never resort to legal coercion to protect any moral good. Resort to legal coercion is itself governed by the rule of reason. As previously noted, focus on the role of legal coercion in matters of public morals may serve to obscure the proper function of political authority in a rightly ordered society. The primary function of political authority is to encourage other responsible organizations of society—familial, educational, religious, fraternal—to exercise their partial roles regarding the protection and promotion of the moral good of the community. Yet, political authority alone has the distinct, specific function of protecting and promoting the moral good of the whole community, and political authority alone has a monopoly of the legitimate means of coercion. When other organizations in society prove inadequate or unwilling to protect and promote the common moral good, political authority may properly invoke its coercive power. The fundamental question is to what extent a democratic and pluralistic polity like the United States should enforce a code of public morals by legal coercion.

What makes this problem so pertinent at present is the general change within the living memory of many of us in the public's code of personal and familial responsibility. Today a large part of the public accepts abortion; tomorrow a large part of the public might accept even involuntary euthanasia. The political scenario on enforcement of public morals seems to follow a pattern: one part of the public urges legislatures to relax or eliminate enforcement of public morals, and another part of the public urges legislatures to maintain or strengthen enforcement. How should citizens react to these political forces? Under what conditions is enforcement of public morals by legal coercion desirable or even possible?

Political authority, of course, is not directly concerned with private or individual morality. The state's legitimate concern, whether in moral or other matters, is restricted to the public and social dimensions of human behavior, to what concerns the community as a whole. Yet, human activities are so interwoven that scarcely any are without some measure of social effect, and so most individual behavior affects the behavior of others. It is the function of practical reason to determine to what degree and with what result individual activities affect public interests. This judgment will depend on how widespread activities are and how bad the effects.

The general purpose of all legally coercive legislation in the field of public morals is to control evils, both physical and moral, that would affect society as a result of citizens acting contrary to moral good. The individual who acts contrary to moral norms incurs the natural sanctions of self-deformity and self-dissatisfaction. So too, the society that disregards communal moral goods reaps natural sanctions of communal deformity and communal dissatisfaction. Organized society exists precisely to enable its members to perfect themselves both physically and morally. To the extent that an organized society works to disorder and frustrate its members morally, it disorders and frustrates itself. As with the individual, so with society, virtue is ultimately its own reward. Political authority, which is specifically entrusted with the protection and promotion of the moral welfare of the community, therefore, should aim to foster social conditions supportive of a sound moral order. This is why even the democratic state, which guarantees absolute freedom in the sphere of religious belief, cannot permit a similar freedom to individuals to act howsoever they choose.

It is difficult to imagine how any rational person could deny the propriety of legal coercion to protect the person and possessions of citizens from violation by other citizens. Likewise, it is difficult to imagine how any rational person could deny the propriety of legal coercion to restrain the behavior of individuals who would corrupt the morals of the young or take unfair advantage of the mentally deficient. But many moderns ask why political authority should be otherwise concerned about the moral behavior of adult citizens enjoying full exercise of their powers to reason and choose. What harm, so the argument goes, can these actions inflict on anyone else? According to many moderns, a "free society" has no rationale to forbid activities to adult citizens on purely moral grounds when no minor or nonconsenting adult is involved.

On the one hand, it should be noted that even morally bad actions which do not immediately affect the person or possessions of others may, when practiced widely, result in harm to others. Drug addicts, for example, may resort to violent crime. But the essential point is that, even absent possible physical harm to other citizens, morally bad activity by individual citizens may pose a specifically moral danger to the whole community. Citizens cannot isolate themselves from the public effects of the moral behavior of individual citizens any more than they can escape the public effects of individual economic behavior. Adult citizens dwell in one and the same society, and young citizens grow up in that society.

On the other hand, legislation that seeks to advance the welfare of the community in one respect may often adversely affect the welfare of the community in another. This is especially true in the case of legislation enforcing public morals. Human law is an exercise of practical reason and so needs to take account not only of the moral objectives to be achieved but also of the social and political costs of using coercion to achieve those objectives. This

means that legislators need to measure whether or not legal enforcement of public morals will do more harm to society than
good. As Jacques Maritain put it, "political hypermoralism is no
better than political amoralism."[9]

One important question to be asked about any law is whether
or not the law can be effectively enforced without recourse to
draconian police methods. A law that cannot effect what it proposes to effect is a silly law. Often enough, political authority
simply cannot enforce public morals, or at least cannot do so
without odious and unacceptable police intrusion into the lives of
ordinary citizens. At best, laws unenforceable in acceptable ways
serve only to declare public policy; at worst, disobedience of unenforceable laws encourages disobedience of all laws. In this respect,
Americans readily recall the legacy of national prohibition. Moreover, ineffectual enforcement of public morals may have other
undesirable social by-products. Illegal abortions, for example, may
be more dangerous to the health or life of women undergoing
them than legally regulated abortions would be.

Another consideration is the pluralistic character of the American polity. Some accommodation of different ethical convictions
about public morals is necessary to that harmony of citizens which
is both a condition and a goal of organized society. Such accommodation, of course, cannot be an absolute consideration. Yet, as a
general principle, legislators should not attempt to enforce public
morals where the community is deeply divided except to protect
paramount public interests such as the lives, liberties, and possessions of citizens. A similar consideration derives from the very
nature of democracy since democracy aims to institutionalize the
maximum freedom of individuals to direct their own affairs. Like
harmony among citizens of different ethical convictions, maximum freedom of individuals cannot be an absolute consideration.
Yet, simply as a political principle, maximum freedom of individuals is a characteristic condition and goal of democratic society.

Lastly, coercive legislation in the area of public expression,
however lofty the moral objectives of such legislation, runs special
risks of inhibiting artistic initiative and scientific progress. More
fearsome still is the political spectre of arbitrary and despotic
power—the very antithesis of rightly ordered political authority—
which censorship raises. This is not to deny that political authority may reasonably restrict some forms of expression (e.g., libel),
but political authority in a democracy should do so only for the

most compelling reasons. The "free marketplace of ideas" is an indispensable condition for citizens of a democratic community to exercise intelligently their political role as decision makers.

Thus, legislators—and ultimately citizens—need to weigh the costs of legal coercion against the benefits to public morals. The case for prohibiting actions that harm other citizens in their persons or possessions is overwhelming, and so political authority necessarily defines and protects the rights of each citizen in relation to other citizens. Political authority at times goes beyond this point to restrict or prohibit actions judged to be seriously harmful to the community even when the actions involve no direct interference with the rights of others, for example, restrictions on drugs, alcohol, prostitution. What restrictions of which activities are judged necessary will depend on how widespread the activities and how bad the results. There is no substitute for practical reason, and Thomas Aquinas summarized the tradition of such reasoning thus: "Human laws do not forbid all vices from which the virtuous abstain but only the more grievous vices from which it is possible for the majority to abstain, and chiefly those that are to the hurt of others, without the prohibition of which human society could not be maintained; thus human law prohibits murder, theft, and such like" (ST 1-2. 62. 2).

The Case of Abortion

Abortion is undoubtedly the prime contemporary problem of public morals in the United States.[10] In nineteenth-century America, there was an overwhelming public consensus against abortion. But the public consensus on the morality of abortion in certain situations (e.g., rape, incest, defective fetuses, danger to the mental health of the mother) had seriously eroded, if not collapsed, by the advent of the 1970s. State legislators might have limitedly liberalized antiabortion statutes to fit the change in public consensus, but they did not generally do so. Nor would limited liberalization of the statutes have satisfied proponents of unlimited abortion. Proponents of abortion, frustrated in the state legislatures, then turned to the federal courts.

In a landmark decision in 1973, *Roe v. Wade*,[11] the U.S. Supreme Court recognized a constitutional right of women to abortions. The Court there, per Justice Harry A. Blackmun, held that the liberty protected by the Fourteenth Amendment includes the

right of women to decide whether or not to have an abortion.
While acknowledging that the right of women to privacy in the
matter of abortion is subject to state regulation, where the state can
demonstrate a compelling public interest, the Court found no such
intcrest on the part of the state to protect fetal life before viability
(the last trimester of pregnancy). This decision managed to com-
bine bad public morals, bad law, and bad politics.[12]

Justice Blackmun and those justices who agreed with him
notwithstanding, the public interest in protecting fetal life during
the entire period of pregnancy is clear and compelling. Fetal life is
actually, not potentially, alive; it is actually, not potentially,
human life at least in the biological sense; it is human life distinct
from that of the mother despite its dependence on the mother.
Human fetal life can be considered "potential" only in the sense
that it will become rationally conscious human life in the course of
development, and human fetal life has the potential for rationally
conscious activity precisely because that life is biologically human
in the first place. Fetal life does not become any more biologically
human or have any more potential for rational consciousness after
viability than before.

Humans, as rational, naturally incline to love whatever is
human in any form. Reason can and should apprehend the human
life of the fetus, whether presently capable of rational consciousness or not, as a moral good simply because that life is biologically
human. As such, fetal life at any stage of its development represents a grave public interest worthy of legal protection. This public interest does not depend at all on whether or when fetal life
constitutes a human person because fetal life itself is a living
human organism in the natural process of developing to a level of
rational consciousness.[13] That is quite enough to constitute a
substantial moral good and to create a substantial public interest
in protecting that good. Indeed, even assuming that the weight of
the argument is against ascribing personhood to the fetus, the
mere possibility that the fetus is a person increases the public
interest in protecting the fetus.

From these perspectives, Justice Blackmun's distinction between the public interest in viable and nonviable fetuses is itself
not viable. The capacity of the fetus to sustain independent life
outside the mother's womb, of course, is of critical importance if
the fetus is forced out of the mother's womb. But the operative
question is whether or not the fetus should be forced from the
mother's womb before viability; given the proper opportunity, the

fetus will become viable. Since the fetus before as well as after viability is a distinct human life at least in the biological sense, it has no less intrinsic value after it reaches the point of viability than it had before. As such, the fetus represents an important public interest worthy of legal protection during the entire period of pregnancy.

Opponents of abortion do their cause no service by claiming that abortion is murder. Murder is the deliberate killing of a human person, and no one has yet satisfactorily demonstrated that the fetus is a human person. Proponents of abortion, on the other hand, are not warranted to conclude from the indemonstrability of the fetus' personhood that the fetus is not human in any sense. On the contrary, the fetus is indisputably a developing human life distinct from the life of the mother whether or not the fetus is a human person. While abortion is not certainly the killing of a human person, it is certainly the killing of a human organism.

No morally responsible citizen can deny the basically human value of fetal life, but morally responsible citizens can and do disagree whether or when abortion is morally justified. Some would argue, as we do, that reason apprehends innocent human life in any form as a moral good, and so that the directly intended killing of a fetus is morally bad. Others would argue consequentially that reason may apprehend other goods as proportionate to the life of the fetus, and so that abortion may be morally justified in cases of rape, incest, severely defective fetuses, and danger to the mental health of the mother if the pregnancy comes to term.

But the moral justification, if any, of individuals choosing abortion in the above mentioned cases does not affect the public interest in protecting fetal life. The aim of law is to protect public interests, not to decide whether or when private citizens act morally in their individual capacities. To protect public interests, legislators may include some morally permissible activities within the sweep of legal prohibitions, as they may exclude some morally impermissible activities from the sweep of legal prohibitions. Not everything legally criminal need be otherwise morally wrong any more than everything morally wrong need be legally criminal. If legal prohibitions sweep too broadly or too narrowly, of course, they may fail to satisy minimum rationality. (As a matter of fact, at the time of *Roe v. Wade* in 1973, almost all American jurisdictions had exempted therapeutic abortions, i.e., abortions when continued pregnancy threatens the life of the mother, from coverage under antiabortion statutes.)

We have argued that the human fetus represents a public interest worthy of legal protection. That argument is a rational argument. No appeal to religious authority or creed is necessary or proper with respect to issues of public morals, and none has been advanced here. Proponents of abortion are wont to claim that opposition to abortion coincides with the tenets of the Roman Catholic religion, and that legal prohibition of abortion would impose Catholic beliefs on those with other or no religious beliefs. On the level of fact, the claim is misleading. Some individual Catholics do not agree with the official Church teaching on abortion, and many non-Catholics do. On the level of argument, the claim is simply irrelevant. The claim does suggest a proper question about whether or not a sufficiently broad consensus exists to support a general prohibition of abortion, but the mere coincidence of antiabortion legislation with Catholic or any other religious belief does not affect the strength or weakness of secular objectives of antiabortion legislation. Moreover, if mere coincidence of public morals policy with the tenets of one or more religions were to constitute some sort of religious establishment, scarcely any public morals policy (e.g., laws on bigamy, gambling, or drinking) could survive constitutional scrutiny.

To say that *Roe v. Wade* failed to accord fetal life the value it deserves as part of the human life process implies neither that state legislators were wise in not limitedly revising nineteenth-century antiabortion statutes before the decision, nor that the decision should not be reversed by constitutional amendment or federal statute. As we pointed out earlier in this chapter, restrictive public morals legislation depends on a broad popular consensus behind it. By 1973, the consensus against abortion had seriously eroded in cases involving rape, incest, severely defective fetuses, and the mental health of the mother, and there is at present no sufficiently broad consensus to sustain and support a general legal prohibition of abortion. Current attempts, therefore, to overturn *Roe v. Wade* seem doomed to fail and unwise from the viewpoint of consensus if the attempts were to succeed. Those who would give legal protection to the life of the fetus need first to revive a braod national consensus against abortion generally before they can hope to enact effective legislation. This means that the primary task of opponents of abortion is one of educating the public about the evil of abortion. It is not likely that such an educational campaign will succeed in the near future, but that is no excuse for failing to make a resolute effort.

The political decision whether or not to prohibit abortion generally is one question; the political decision whether or not publicly to fund abortions for indigent women is another. While it may be morally right public policy not to prohibit abortion in the absence of a sufficiently broad consensus, it cannot be morally right positively to promote abortion if abortion is morally wrong. Public funding of abortions for indigent women is precisely a case of the latter. Therefore, legislators who realize that abortion is morally wrong should not vote specifically to fund abortions, nor should governors or presidents morally opposed to abortion sign specific legislation funding abortions.[14] To do so would be morally equivalent to lawmakers themselves performing the abortions they vote to fund.[15] This is undoubtedly a hard political issue for lawmakers morally opposed to abortion with large constituencies in favor of abortions for indigent women.

Some suggest that the failure of the public to fund abortions for indigent women leads to an unfair two-tier system whereby relatively affluent women can afford to choose abortions while poor women cannot. This is surely an anomaly and an inequity of sorts. But the inequity does not consist in the fact that poor women are denied the means to choose abortions. Rather, the inequity consists in the fact that relatively affluent women are permitted to choose abortion. The moral way to remedy the inequity is to deny to all the legal right to choose abortions, not to fund the poor to do moral wrong.

Administrators of public funds appropriated for the abortions of indigent women are in a different position from that of lawmakers. Administrators are only responsible for distributing funds according to the lawmakers' disposition. Their cooperation with the abortions funded is limited to the execution of a policy already decided on by lawmakers and over which the administrators have no control. Moreover, if one administrator were to resign rather than administer the law, a replacement would undoubtedly be found to do so. Administrators of laws publicly funding abortions are not called upon to approve the laws, nor are they as individuals indispensable to carrying them out.

Notes

1. The translation is that of Martin Ostwald, *Nicomachean Ethics* (Indianapolis: Bobbs-Merrill, 1962), p. 34.

2. The Supreme Court has interpreted the religious establishment clause to circumscribe severely the role that public schools may play in the moral education of students. *Abington v. Schempp*, 374 U.S. 203, 83 S.Ct. 1560, 10 L.Ed.2d 844

(1963). On the constitutional possibility of religious and/or moral instruction in public schools on an elective basis with adequate secular alternatives, see my article, "The Dilemma of Religious Instruction and Public Schools," *The Catholic Lawyer* 10 (Winter 1964):42–54, 82.

3. The religious establishment clause enjoins the government from directly aiding religion. However, the government can and does indirectly aid religion when the government protects citizens' exercise of religion, as in the case of providing military chaplains.

4. John C. Murray, S.J., *We Hold These Truths* (New York: Sheed and Ward, 1960).

5. Sir Ernest Barker, *Traditions of Civility* (Cambridge, Eng.: Cambridge University Press, 1948).

6. Walter Lippmann, *The Public Philosophy* (Boston: Little, Brown, 1955).

7. *Minersville v. Gobitis*, 310 U.S. 586, 596, 60 S.Ct. 1010, 1014, 84 L.E. 1375, 1380 (1940).

8. A subsidiary, procedural question concerns the propriety of religious leaders attempting to influence public policy when they speak out on moral issues. The First Amendment guarantees the right of religious leaders to speak freely as much as it protects the right of other citizens to do so, and that Amendment guarantees the right of religious leaders to exhort to moral practice as much as their right to preach religious doctrine. The use of tax-exempt property for openly partisan purposes, however, would seem inconsistent with the purposes of exemption (religious worship and associated activities) and, indeed, to accord a privileged position in the political process to religious leaders.

9. Jacques Maritain, *Man and the State* (Chicago: University of Chicago Press, 1951), pp. 61–62.

10. The only abortion considered in this section is direct abortion, i.e., the deliberately intended expulsion of a nonviable embryo or fetus from the mother's womb. It is also assumed here that direct abortion involves an intention to kill the embryo or fetus, an assumption verified in almost all current abortions. Indirect abortions, as in the case of removal of a cancerous womb with an embryo or fetus, can be morally justified on the principle of double effect. So, too, in my opinion, can direct abortions which do not involve an intention to kill the embryo or fetus, as in the case of aborting a nonviable fetus whose development threatens the life of the mother. Presupposing use of conservative means when a direct abortion is performed to save the life of the mother, the aborted fetus will die because it is nonviable, not because participants desire or intend its death. As indicated in Chapter 1, therapeutic abortions to save the life of the mother are analogous to cases of repelling life-threatening aggression by unwitting adult agents. But cf. *Casti Connubii* (Denzinger-Schönmetzger, n. 3720).

11. *Roe v. Wade*, 410 U.S. 113, 93 S.Ct. 705, 35 L.Ed.2d 147 (1973).

12. The text concentrates on public morals aspects of *Roe v. Wade*. As to why that decision represents bad law and bad politics, consider the following points. The Court substituted its own judgment about the relative rights of pregnant women and fetuses for that of state legislatures but could cite no explicit constitutional provision for the Court's authority to do so. The Court relied on the due process clause of the Fourteenth Amendment. The Court normally asks states only to show minimally rational justification for restrictions on individual citizens' liberties. In the case of liberties connected with democratic processes, like free speech, the Court insists that states demonstrate substantial or compelling reasons for restrictions. The Court in *Roe v. Wade* singled out women's rights to abortions for similar preferential treatment, although such alleged rights have no relationship to the political process. In short, the Court read its own views on the substantive question of abortion into the constitutional requirement of due

procedure. Moreover, the Court interpreted the due process clause of the Fourteenth Amendment to protect women's rights to abortions during the first two trimesters of pregnancy in spite of the fact that 36 state or territorial legislatures prohibited such abortions at the time the Fourteenth Amendment was adopted in 1868.

As to the bad politics of the decision, the Court apparently hoped thereby to remove abortion from the political arena. Subsequent events demonstrate that the Court badly underestimated the strength and depth of a large segment of the public's opposition to general abortion.

13. Aquinas, following Aristotle, did not regard impregnated ova as human persons before development deemed suitable for the presence of a human soul (*De veritate catholical fidei* 2. 89). In a similar vein, many moderns, identifying personality with consciousness, do not recognize embryos or fetuses as human persons. In addition, the phenomena of the twinning and conjunction of zygotes pose formidable difficulties for any theory espousing the presence of human personality before implantation.

14. Lawmakers, however, may morally vote for or sign omnibus appropriation bills which include funds for abortion on the principle of double effect. Lawmakers may reasonably conclude that the loss of worthwhile public projects were the omnibus bills to fail constitutes a greater evil than the assistance to abortion were the omnibus bills to become law. On the same logic, citizens may vote for candidates who support public funding of abortions if they conclude that the loss of worthwhile public projects were the proabortion candidates to be defeated constitutes a greater evil than the assistance to abortion were the proabortion candidates to be elected.

15. To avoid greater evil, legislators have the option of abstaining from votes on specific bills funding abortion, and governors and presidents have the option of allowing specific bills funding abortion to become law without their signature.

5

The Justice of Economic Institutions and Distributions

Most, if not all, major political theorists have considered the relationship of property to organized society and the good life. Property links humans to the physical world and to one another. On the individual level, humans depend on material goods to live and develop themselves as persons. On the interpersonal level, property joins humans to one another in the production and consumption of material goods. The relation of property to persons is a central concern of justice because property relates things to persons, one person to another, and individual persons to social groups. And because property relations are a concern of justice, they are a concern of organized society. There are three basic views about the relationship of property to the good life: the classical liberal view, the communist view, and the Aristotelian-Thomistic view.

Classical liberalism looks upon the preservation of private property as the highest goal of organized society apart from protection of life and limb. As Locke puts it, the "chief and great end . . . of men's uniting into commonwealths and putting themselves under government is the preservation of their property" (TG 9. 124).[1] Communism takes the diametrically opposite position, that is, private property is the source of all conflict among humans and dehumanizes them. Aristotle and Aquinas defend the position that the use of property is necessary for human development, and that private ownership of property is a relatively better means than common ownership to achieve that development.

The contemporary liberal democratic view of property, regime, and human good is a hybrid view of classical liberals, on the one hand, and that of Aristotle and Aquinas, on the other. C. V. Macpherson has described the liberal democratic view as one of "possessive individualism," a view which would maximize both individual utilities and the development of human powers.[2] In fact, as Macpherson argues, the ideal of maximizing individual utilities has swallowed up the ideal of maximizing human powers; the ideal of infinite appropriation and consumption has left no room for the ideal of human development.[3] Moreover, possessive individuals can maximize their utilities only at the cost of other individuals' ability to develop their human potential.

In Chapter 3, we dealt generally with the justice of political institutions. We return to that topic now in the context of the body politic's economic institutions and the economic distributions which result. One part of the justice of economic institutions concerns the rights which individuals and groups enjoy and the duties which they owe to one another. The state, as the agency of the body politic responsible for specifying the common good, will define such things as the rights and duties of management and organized labor as well as regulate industrial conflict when it occurs. This part of the virtue of justice Aquinas calls commutative justice (ST 2-2. 61. 1). The other part of the justice of economic institutions concerns the proper distribution of benefits to members of society in return for their contributions to the common good. As we shall see, Aquinas would have the state, as the chief instrument of the body politic, regulate the distribution of economic benefits so that all members of society have sufficient goods for their human development. This part of the virtue of justice Aquinas calls distributive justice (ST 2-2, 61. 1). The two parts of justice are closely interrelated; the definition of the rights and duties of citizens in property relations will necessarily affect the distribution of economic benefits. But the two parts of justice are conceptually distinct.

Questions about the rightness of economic institutions and distributions arise only in connection with market societies. In preliterate, premarket societies, individuals or kin groups responsible for the production of goods also trade the goods, and so producers decide for themselves what to do with their surplus product.[4] But in market societies, the transfer of goods from producers to consumers is effected by middlemen. This raises questions about the justice of relations between producers and middlemen and between middlemen and consumers. It also raises questions about the resulting distribution of economic benefits to producers and middlemen because, in the absence of regulation, the marketplace will inevitably result in unequal distribution of economic benefits.[5]

At the outset, it is important to note what is not involved in questions about the justice of economic institutions and distributions. First, poverty is not necessarily the product of economic injustice; poverty may indeed be the result of the way in which society is economically organized, but that need not be the case. Individuals may be poor through their own fault, as will be the result if they are lazy or improvident. Or acts of God may impoverish individuals, as in cases of sickness and natural disasters.[6] Indeed, entire communities may be poor because natural resources are lacking, as is the case with sub-Saharan tribes.[7] In none of these cases can we speak of a just or unjust organization of economic institutions or of a just or unjust distribution of economic benefits. Conversely, economic injustice need not necessarily result in dire poverty for some or most members of society, and such economic injustice will remain so even though of less than dire proportions.

Second, the justice of economic institutions and distributions is not a matter of charity. On the one hand, there are moral obligations beyond those of justice. The virtue of charity requires individual citizens to help those in need to a reasonable extent despite the fact that the donor bears no responsibility for the beneficiary's situation of need. And the virtue of charity requires organized society itself to help those in need to a reasonable extent despite the fact that the beneficiaries themselves are responsible for their situation of need. On the other hand, individuals and groups may have moral claims on organized society on the basis of economic justice beyond appeals to human compassion. Charity is a duty of the donor, not a right of the beneficiary, while justice is a matter of reciprocal rights and duties.

Third, economic justice is not only a matter of the duties a nationally organized society owes to its own citizens but also a matter of the duties economically developed nations owe to underdeveloped nations. Nonetheless, there is at least one significant difference between duties of economic justice within nation states and duties of economic justice among nation states: no internationally organized society exists to make authoritative decisions allocating economic burdens and benefits among nations. The United Nations is only an association of nation states and depends on the consent and cooperation of member states to carry out policies. As a result of this lack of international centralization, it will be difficult both to determine the specific duties of individual nations one to another and to fulfill those duties when determined.

Our starting point will be Aquinas' views on the just economic organization of society. We shall then attempt to evaluate the adequacy and relevance of his views for the modern world. In the light of that evaluation, we shall develop general principles concerning the justice of economic institutions and distributions and apply the principles to class, capitalist, communist, and welfare capitalist economic systems within nation states. Lastly, we shall examine the responsibilities of developed nations to underdeveloped nations.

Aquinas on Property[8]

As we have seen in Chapters 1 and 2, human good consists of activities of reason and other activities according to reason, and this human good can only be achieved in association and cooperation with others. The purpose of property is not only to sustain human existence but also to make possible human development. Material goods are a necessary means and a necessary condition— but only a means and only a condition—for the development of the potential of the human person for knowing and loving. Accordingly, Aquinas rejects any conception of human development as a process of maximizing subjectively desired material utilities; to act as if the possession and consumption of material goods were the goal of human existence would be to follow an inhuman, indeed an antihuman, path of development. Ordering material goods to the development of human personality is both the proper goal of individuals and of organized society.

Assuming that the purpose of property is to support and assist human development, the question naturally arises as to what amount of property is consistent with that objective. Aquinas answers the question thus: self-sufficiency in material goods is "that [amount] which . . . makes life desirable and free from wants" (CE 1. 9. 114).[9] Self-sufficiency in material goods is inconsistent with both superabundance (CE 10. 13. 2128) and boundless desire for material goods (CE 1. 9. 116). The private acquisition of material goods for a self-sufficient human life, however, is ordered to the good of the community, especially the moral good of the community, and political authority has the responsibility, among other things, to secure for members of the community that sufficiency of material goods necessary for living a proper human life (DR 1. 14-15). Political authority is required not only to discipline wrongdoers but also and especially to direct all members of the community toward the goal of proper human development. Laws are necessary to coerce the unvirtuous to live virtuously and to guide the virtuous in the path of virtue (CE 10. 14. 2153).

If property is a necessary means to genuine human development, property should be accessible to all humans, and this invites a question about the rights of individuals to possess property as their own. Material things are made for human use, and humans, as rational, are capable of producing and using things for their common self-development (ST 2-2. 66. 1). Nonetheless, it is just that individuals have the right to acquire and dispose of things as their own (ST 2-2. 66. 2). The natural law prescribes only that the use of goods be common, not that the possession of goods be common. Therefore, social convention expressed in civil law can justly assign to individuals the possession of goods to dispose of as their own (ST 2-2. 66. 2. ad 1; cf. 1-2. 94. 5. ad 3).

Aquinas assigns three reasons for the relative advantages of private over common ownership (ST 2-2. 66. 2; cf. 1. 98. 2. ad 3). Whatever the case for common ownership in an ideal world composed of completely virtuous citizens, common ownership in the actual world of less than completely virtuous citizens would invite greater conflict and injustice than would private possession. First, with private ownership, each human will be more careful to take care of property; if things were possessed in common, humans would tend to shrink from work and leave to others the care of property. Second, with private ownership, management of property will be more orderly since each human will know exactly what property he or she is responsible for; if things were possessed in common, humans would be confused about their responsibilities for the care of property. Third, with private ownership, the community will be more peaceful since each human will know what is his or her property; if things were possessed in common, there would be frequent quarrels about the division of goods. In his *Commentary on the Politics* of Aristotle, Aquinas adds another reason why organized society should institute a system of private rather than common ownership of property. If goods were possessed in common, those who did not work or worked less would be rewarded equally with those who did work or worked more; this would violate justice because those who do unequal work should not receive equal reward with those who do (CP 2. 5). Lastly, there is a natural link between individuals and what they produce, and so land more properly belongs to the individual who cultivates it than to others who do not (CP 2. 4).

A system of private possession stimulates production better than common possession, but the principles of self-sufficiency and common use subordinate material production to human development and just distribution, respectively. Unless the aquisition of material goods is limited to what is necessary for human development, the quest for material goods will become disoriented. Similarly, unless private possession is subordinated to common use, an unjust distribution of material goods will result. The right to common use means minimally that humans without the material goods necessary for subsistence have the right to use superfluous material goods in the private possession of others. For this reason, Aquinas denies that the forcible taking of things necessary for human subsistence from the superfluity of others constitutes theft (ST 2-2. 67. 7). More broadly, the right of individuals to produce and exchange material goods should not exclude other individuals from use of those material goods necessary for their human development. In classical liberal and liberal democratic theories of private property, on the other hand, individuals have absolute rights of ownership over the things they produce or exchange to the exclusion of other individuals and their needs.

Property relations involve justice, and the essence of justice is a "kind of equality" (ST 2-2. 57. 1; cf. 2-2. 58. 2). Like all virtue, justice is a mean, in this case a mean between the external thing and the external person (ST 2-2. 58. 10; 2-2. 59. 2). Justice as equality implies for property relations that every human should have an equal opportunity to obtain sufficient property for his or her human development.[10] Principles of commutative justice govern exchanges of goods and services, and parties to the exchanges have a claim in justice to an equal return (ST 2-2. 61. 3), indeed to an arithmetically equal return (ST 2-2. 61. 2). The value of what is exchanged is ultimately determined by the suitability of the goods for human use (ST 2-2. 71. 1).

Aquinas distinguishes between the proper or intrinsic value of goods and their common or exchange value (CP 1. 7). Aquinas cites shoes as an example to illustrate the distinction. The proper use of shoes is as covering for the feet. But shoes have an alternate use when exchanged for something else (e.g., food). The use of goods in exchange for other goods is valid as long as the goods exchanged are traded according to their proper or intrinsic value. The market price of goods may or may not reflect the intrinsic value of the goods. For example, the law of supply and demand of the marketplace may raise the market price of goods beyond their intrinsic value. In such a case, sellers should not seek to profit from the need of buyers (ST 2-2. 77. 1). To do so would be to seek profit for profit's sake, and this Aquinas will not permit; exchanges should be transacted only for the sake of proper human development (ST 2-2. 77. 4. ad 2).

An organized society not only regulates exchanges of property but also how individuals are rewarded for what they contribute to the good of the community. Distributive justice requires that there be a geometrical or proportional, not an arithmetic or absolute, equality in the rewards individuals receive for their contributions to the common good (ST 2-2. 63. 2). Let us assume that professors, for example, receive twice as much salary as secretaries, and university presidents twice as much salary as professors. If professors contribute twice as much as secretaries to the common good of educating citizens, and university presidents twice as much as professors, there will be a proportional equality in the distribution of rewards to the three groups. How is the value of the individual's contribution to the common good determined? Ideally, the intrinsic quality of contributions should determine the hierarchy of social values, but one society may esteem certain virtues more than other societies and distribute rewards accordingly. Thus, scholars may make an intrinsically greater contribution to the human development of the community than businessmen, but a society intent on moderate economic development might choose to reward businessmen more than scholars. For a society to sanction rewards for individuals simply because of their connection with those in a position to award benefits, however, would be beyond the reach of any proportionality (ST 2-2. 63. 3).

Exercise of the virtue of economic justice involves exercise of the virtue of moderation. Individuals cannot act justly toward others in economic matters if they appropriate and use an amount of wealth more than what is necessary for their human development. To seek to acquire more material goods than necessary for human development would be to make material goods ends in themselves to the exclusion of the well-being of others (CP 1. 8). Profit itself is not incompatible with the virtue of moderation, but profit for profit's sake, that is, profit for the sake of individual appropriation without regard for social purposes or consequences, is incompatible. Profit may be sought for the sake of such things as the reward of one's labor, the good of the community, aid to the poor, or provision for one's future (ST 2-2. 72. 4. ad 2). Property produced by one's own labor is not immoderately appropriated (ST 2-2. 55. 6. ad 2). Nor is it necessarily immoderate to consume more goods in order to maintain a higher social position than another would need to maintain a lower social position (ST 2-2. 32. 6; 2-2. 141. 5).

As exercise of the virtue of economic justice requires exercise of the virtue of moderation, so exercise of the virtue of moderation requires exercise of the virtue of liberality (CP 1. 4). Liberality corrects the tendency to regard acquisition of material goods as the end of human existence and disposes individuals to act toward others as human nature intends. An antidote to "possessive individualism," liberality diverts individuals from love of money for its own sake and inclines them to use money properly for their own human development and that of others (ST 2-2. 117. 6).

When humans desire to acquire property without limit, property no longer is regarded as a means for human development but instead becomes an end in itself (ST 2-2. 118. 2). This elevation of property from a means into an end proceeds from an unnatural appetite (ST 1-2. 30. 4), and it is money that makes it possible for humans to desire to possess material goods without limit (CP 1. 9). In property transactions, money serves as a medium of exchange (CP 1. 7). Because money is capable of measuring the value of commodities, it becomes possible to exchange commodities of unequal value. The same capacity of money to measure all material goods makes it possible for disordered humans to desire to possess property without limit by means of money. One cannot, of course, possess all property, but one can desire to do so. To use money for the sake of acquiring more money is to make money an end in itself without reference to human development. In a money-oriented society, money becomes the most important thing to possess (ST 2-2. 118. 7. ad 1), enables its possessors to possess property without limit (ST 2-2. 118. 7. ad 2), becomes an object of idolatry (ST 2-2. 118. 7. ad 4), and gives a false sense of self-sufficiency (ST 2-2. 118. 7).

Avarice, the immoderate desire for material possessions, sooner or later gives rise to all other social evils and, in that sense, can be called the root of all evil (ST 2-2. 119. 2. ad 1). Moreover, in Aquinas's view, for one person to acquire superfluous wealth, that is, more wealth than is sufficient for human development, necessarily involves another person having insufficient wealth for human development: "one man cannot overabound in external riches without another man lacking them, for temporal goods cannot be possessed by many at the same time" (ST 2-2. 118. 1. ad 2). Nor can individuals acquire superfluous wealth without using force and fraud (ST 2-2. 118. 3). From avarice, a full panoply of evils directly stems: dissatisfaction, violence, deceit, lies, perjury, betrayal, greed, and inhuman sensitivity (ST 1-2. 61. 2; 2-2. 55. 6. ad 8). The avaricious appetite is insatiable, and individuals with such an appetite will seek to compensate for their spiritual emptiness by acquiring more and more material possessions, to maximize material utilities to compensate for failure to develop their human potential (ST 2-2. 118. 5. ad 3). Aquinas's conclusions about the purposes of property can be summarized as follows: property is a necessary instrument of human development, and property should be used moderately and justly. Of the two evils of surplus possessions and greed for material possessions, the latter is the greater obstacle to human development (CP 2. 9).

Although the cultivation of virtue is more important for human development than the legal regulation of property uses, the latter is nonetheless necessary. Humans live in association with other humans, and that society is organized. The state gives direction to human development, orders property relations and property uses to that end, and coerces citizens to observe rules about property (DR 1. 1). This view contrasts both with that of Adam Smith and that of Karl Marx. Smith did not think the state necessary with respect to property distributions, and Marx thought the state an instrument of property maldistribution which would become unnecessary in a perfectly reconstructed society. For Aquinas, the state is necessary to achieve just property relations and just property uses, can substantially achieve that object, and so will never wither away.

According to Aquinas, individuals should have the right to possess property as their own, and they have the duty to put such property to common use. The obvious tension between private ownership and common use is to be resolved by the state as a "good legislator" (CP 2. 4). The state alone has the specific responsibility for the good of the community. Exercise of the responsibility for the common good in matters of property relations and property uses will translate into legislation to correct the antisocial consequences of acquisitive activity by individuals. The state decides for the community which acquisitive activities are socially harmful and which are not.

The state may even decide to tolerate a morally bad activity like usury on the principle of double effect. Usury, in the medieval sense, is any lending of money to another at interest. Aquinas regarded usury as morally wrong because money has no intrinsic value beyond its value as metal or art and, therefore, serves principally as a medium of exchange. To charge another for something that has no intrinsic value would be to exact something for nothing, to seek profit for profit's sake (ST 2-2. 78. 1). Aquinas gives two reasons why a good legislator might decide to tolerate usury: (1) many citizens lack enough virtue to avoid the practice of usury; (2) the community would lose many advantages if usury were prohibited (ST 2-2. 78. 1. ad 3; cf. CP 2. 4). The "many advantages" of usury presumably include promoting commerce.

Although Aquinas judged an agricultural economy more conducive to human development than commerce, he accepted the legitimacy of a modest amount of commercial activity (DR 2. 3).[11] Aquinas sees two positive values in commerce. First, since no organized national society can be self-sufficient, it will need to trade with other organized national societies. Second, for an organized national society not to trade its surplus commodities in return for the surplus commodities of other organized national societies would serve no rational purpose. Note that Aquinas does not say that desire for profit justifies commerce, but rather that commerce avoids waste and satisfies needs.

Aquinas did not envision for the state any role of producer or stimulator of production: "It is not the proper or immediate duty of government to produce or acquire food, but it is a proper duty to dispense . . . those things according to need" (CP 1. 8).[12] This is to say, Aquinas envisions for the state a role of regulating economic activity so as to assure a just distribution of goods, not a role of directly producing commodities or managing their production. Aquinas would be unsympathetic to state socialism and to state capitalism because both would have the government do what individual citizens and subordinate social units should do for themselves. (State capitalism, in Aquinas's judgment, would have the additional defect of stimulating private individuals and groups to lend money at interest and so of promoting profit for profit's sake.) From the perspective of individual citizens, virtue should motivate them to produce enough goods for their self-sufficiency. From the perspective of the state, the common good should govern regulation of economic activity to insure a just distribution of goods. Aquinas would formulate the principles of economic justice for individuals and the state thus: (1) individuals should contribute to the common good according to their ability, and they should receive rewards proportionate to their virtuous contributions; (2) the state should regulate economic activities to insure that individuals do receive rewards proportionate to their contributions.

Critique of Aquinas on Property

Aquinas situates economic good within the framework of human good and subordinates the former to the latter. In that context, Aquinas is surely correct that the desire for unlimited possession of material goods is inconsistent with genuine human development. But the question remains whether or not he is correct in limiting the amount of material goods consistent with human development to mere sufficiency in the sense of freedom from want. There would seem to be no a priori reason why material comfort beyond freedom from want need be incompatible with human development, and there may be a posteriori reasons why such comfort could be more compatible. Typewriters and word processors, for example, assist the intellectual life at the same time that they make material life more comfortable.

When Aquinas limits the possession of material goods to the amount necessary for human development, he seems to be principally concerned about the human development of the upper classes of medieval society. Aquinas approves only limited commerce, and the economic consequence of limited commerce is necessarily limited products to distribute to the lowest classes in society. Serfs in medieval society might have been able to achieve moral virtue, that is, to act according to reason, but they hardly had sufficient material resources to exercise intellectual virtue, that is, to engage in activities of reason. Moreover, limited production will freeze most individuals into fixed economic positions in society. Serfs not only had no opportunity to choose their own occupations, but their children had no opportunity to choose theirs (except, to an extent, priesthood or knighthood). The population explosion of modern times will require vastly expanded economic production if most individuals are to have sufficient material goods for their proper human development. In modern societies, growth of national product is essential if all members of society are to have sufficient material goods for their human development.

Aquinas's claim for the communal dimension of property is a necessary corrective for those who would view property only in terms of individual acquisition. However, capitalist theory claims that it best satisfies the overall common good, and that theory should be examined in detail. Similarly, Aquinas's claim that private, as opposed to common, possession of material goods better accommodates human development should be reexamined in the light of the Marxist critique.

Aquinas predicates commutative justice in property relations on the intrinsic value, that is, the proper use-value, of the commodity traded. Undoubtedly, a decent meal has objectively more intrinsic value than junk food, and a pair of shoes objectively more intrinsic value than an electric pencil sharpener. But the use-value of at least some commodities also involves subjective estimates of utilities, and, to the extent that citizens recognize utilities in an objective framework, the marketplace will measure the fair price of commodities.

Aquinas defined distributive justice as that distribution which allots rewards in proportion to contributions to the common good. Class, capitalist, and communist economic systems all purport to satisfy distributive justice so defined, and their respective claims will have to be examined in greater detail. For the moment, it is important to stress the fact that Aquinas links economic rewards in society not to the marketplace value of labor as a commodity but to the virtue of citizens. This linkage to virtue means at least that workers should be rewarded according to their personal worth, not simply their market worth, if the market worth falls below the minimum necessary to support workers' human development. Whether or not just distribution requires more than that minimum or even absolutely equal distribution remains to be considered.

According to Aquinas, money has no intrinsic value apart from its metallic or artistic uses. Modern capitalism, on the other hand, treats money as a commodity in its own right and not merely a medium of exchange; its principal intrinsic value, as opposed to its exchange value, is to promote manufacture and commerce. Nor do lenders earning interest on loans get something for nothing. Lenders take the risk of losing their money, and lenders lose the opportunity to use the money lent for their own business or personal consumption. Aquinas does not consider the lenders' risk of

losing money, but he does recognize that lenders may lose opportunities to use the money lent for themselves, and that they are
entitled to recompense for those lost business opportunities (ST 2–
2. 78. 2. ad 1). If lenders are entitled to be compensated for their
lost business opportunities, the capitalist theorist may ask why
lenders should not be compensated for their lost opportunities to
lend money at a profit.

Aquinas condemns profit for profit's sake on the basis of his
argument against the intrinsic value of money. But if Aquinas's
theory of money can be rejected, then moderate profit for profit's
sake should be acceptable on basic Thomistic principles as long
as profit serves social purposes. As previously indicated, Aquinas
admits that a regime may tolerate the lending of money at interest
in order to preserve the advantages the practice brings to many.
Aquinas seems thereby to recognize not only the advantages that
interest-taking brings to the economic life of the community but
also a distinction between subjective intentions of lenders and objective consequences of interest-bearing loans, which consequences
are not dissimilar to those produced by Adam Smith's "invisible
hand."

Aquinas distinguishes lending money at interest from providing money as a business partner. In the former case, the lender
transfers ownership of money to the borrower; in the latter case,
the provider of money does not relinquish ownership. The money
provided by a partner remains his or hers, and a merchant or
artisan uses it at the partner's risk. Aquinas would allow the
partner to share in the profits of the enterprise (ST 2–2. 78. 2. ad 5).
This practice, which was not deemed profit for profit's sake, was
one of the principal ways medieval businessmen circumvented the
Church's strictures against usury.[13] But it is difficult to see any
practical difference between partners providing money in the hope
of profit and lenders lending money at interest. The lender also
risks loss of his or her money if the enterprise fails, and a partner
contributes no more to the enterprise than a lender. The distinction between partners and lenders seems to exalt form over substance.

Aquinas would have us believe that to seek profit for profit's
sake would necessarily imply a desire to possess material goods
without limit. But does it? To seek profit for profit's sake would
certainly imply a desire for more material possessions, but "more"
is not the same as "unlimited"; a lender might seek to make only a
moderate profit. Aquinas would also have us believe that for some
citizens to have more material goods than are sufficient for human

development necessarily implies that other citizens will not have sufficient material goods for their human development. But that assumption would only be true if the sum total of available material goods is no greater than the sum total sufficient for the human development of citizens. It seems possible in economically developed societies to produce more material goods than those necessary to free all citizens from want. Aquinas would also have us believe that the acquisition of superabundant material goods necessarily involves the use of force and fraud. But while the desire for more material goods than those sufficient for human development may lead some to use force or fraud, there seems to be no a priori reason to suppose that most citizens cannot act justly to increase their material possessions.

Lastly, Aquinas would have the state regulate property relations and property uses to achieve a just distribution of material goods but permit the state neither directly to produce material goods or services nor indirectly to stimulate their production. Aquinas thought that the production of goods and services properly belonged to individual citizens and voluntary groups, and that virtuous citizens could and would produce adequate goods and services without any intervention by the state. In developed societies, however, individuals and voluntary associations may not produce enough material goods and services, or at least certain kinds of goods and services, for the economic and social well-being of the community, and the state will have the responsibility to stimulate their production.

Some of Aquinas's views on property seem to flow necessarily from his concept of human nature and human good: economic good is subordinated to human good, economic good should be sought in moderation, and material goods are for the benefit of the entire human community. Other of Aquinas's views seem not to flow necessarily from his concept of human nature. Aquinas argues only that private property is consonant with human nature, and Aquinas's views on money and interest-taking seem culture-bound. Therefore, it is reasonable to reexamine the relative justice of different economic systems in the modern context.

The Justice of Class, Capitalist, and Communist Systems

In Chapter 3, we argued that institutional political justice requires that citizens enjoy as much freedom as is compatible with the common good, that citizens be treated equally under a rule of law, and that inequalities in organized society need to be justified in terms of the well-being of the community. These principles apply to the institutional justice of economic systems; economic systems should leave citizens free to decide what they will produce and consume as far as that is compatible with the common good, citizens' economic freedom should not be restricted on the basis of personal characteristics like race, religion, or sex, and economic inequalities need to be justified by objective factors related to the good of the community. But the justice of economic systems in modern societies involves three special distributional considerations: the effect of the systems on the gross product of goods and services, the general acceptability of distributions, and the compatibility of distributions with the human development of all contributors.

Market economies take traditional and modern forms;[14] the traditional form of market economy is an ascriptive class system, and modern forms of market economies range from pure capitalism to pure communism. We shall first consider the justice of the economic institutions and distributions associated with pure class, capitalist, and communist systems. We shall consider subsequently the justice of the economic institutions and distributions associated with welfare capitalism, a modern form of market economy which mixes elements from the pure capitalist and communist models.[15] We are not seeking to find an economic system which is perfectly just; rather, we are seeking to determine which economic system or systems come closest to realizing ideals of justice.

An ascriptive class system assigns economic burdens and benefits on the basis of heredity. Medieval society in full flower is a classic example of such a system. Most members of medieval society were ascribed at birth to a particular economic and social position. The children of peasants became peasants, and the children of nobles succeeded to positions of privilege. There were a few avenues of upward mobility for males (e.g., knighthood, priesthood), but the children of peasants and artisans could not easily satisfy the relatively high prerequisites of those limited opportunities. At the heart of the medieval economic system was agriculture. Surplus agricultural production supported nobles owning the land and warriors protecting it. Each class was theoretically rewarded for its contribution to the product: peasants for their labor, nobles for administering production and distribution, and warriors for protecting producers and the product. Similarly, medieval guilds divided the fruits of craft labor between merchants and craftsmen by fixing prices and wages. And universities trained lawyers and clerics to regulate economic and social conflict. In short, the medieval principle of economic and social justice can be summed up as follows: from each according to his or her class, to each according to his or her class.

Enlightenment philosophers, however much they exalted individuals over the social bonds joining them, rightly stressed the values of personal freedom and equality. Absent justification in terms of the common good, ascriptive class systems fail to satisfy the institutional principles of maximum freedom and equal treatment. As to freedom, most individuals in class systems have no freedom to choose their occupations, and peasants have no say in the reward they receive for their labor. As to equal treatment, ascriptive class systems by definition treat citizens unequally. As to distributional goals, class systems provide little or no incentive to increase production or make production more efficient, citizens are unlikely to accept unequal distributions on the basis of ascriptive classes if an alternate system of distribution offers more benefits, and those at the bottom of the scale of rewards are unlikely to receive compensation minimally compatible with their human development. Special circumstances in the Middle Ages might have justified the feudal system for a time as the least bad way of organizing society for the benefit of all, but such circumstances could not permanently justify the system. Indeed, principles of institutional and distributional justice require not only that ascriptive class systems be opportunely dismantled, but also that the conditions of the times which make the systems acceptable be changed.

Medieval society attempted to justify vastly superior rewards to noble and warrior classes of the feudal system on two grounds. First, the warrior class was necessary to protect producers from attack. Second, the noble class of landowners was necessary to supervise both defense of the product and its marketing. But whatever merit that justification of medieval society may have, there is far less justification of the feudal system which exists today in pockets of Latin Europe and much of Latin America. Central governments have succeeded the diffuse medieval political system, and so landowning classes have no political or military justification for their privileges in most parts of the modern world. Second, landowners are less necessary to administering production and distribution in most parts of the modern world than in the medieval setting. Third, and critical from the perspective of social acceptability, landowners' efforts to maximize profits from large-scale agriculture have undermined whatever paternal role toward peasants landowners may once have played and so the legitimacy peasants may once have given to the class system.

As opposed to traditional economic and social class systems, modern capitalism claims to reward individuals economically according to the value of what they produce, and laissez-faire theories boast that pure capitalism leaves all individuals free to compete in the marketplace on equal terms. Since capitalist systems have historically operated in a framework of political as well as economic freedom, they seem to satisfy the principle of freedom very well; individuals seem to have both political freedom and the economic freedom to choose their occupation, to negotiate the price of their goods and services, and to select what they will consume. But the freedom of those less economically privileged to compete in the marketplace on equal terms with those more privileged is illusory. Successful capitalists have overwhelming advantages in future competition. They have capital, and they have access to more capital from relatives and friends. Second, successful capitalists can provide their children with superior educational opportunities and environment. Third, middle- and upper-class parents are able to transmit entrepreneurial values and motivation to their children better than working- and lower-class parents. Lastly, individuals, through no fault of their own, may come to the marketplace with physical or psychological disabilities.

Concerning the economic distributions that capitalism produces, the capitalist model may seem at first sight to have no theory of social justice at all. Since laissez-faire theorists argue primarily for the freedom of individual entrepreneurs from social controls, they may seem totally to divorce the economic good of the individual entrepreneur from the good of the community. Such an interpretation would too summarily dismiss the "invisible hand" part of theory. Leave entrepreneurs to themselves, so the fuller theory runs, and the results will ultimately benefit most or all members of society.

Capitalist theorists claim that a free market economy will inevitably stimulate production and make production more efficient. The historical record supports that claim. Free market economies, based on the profit motive, have vastly stimulated production in the West over the last several centuries, competition has led to vastly more efficient production, and citizens of the West generally enjoy vastly improved standards of living. The economic virtues of capitalism can be easily exaggerated, and its social vices easily ignored.

Although capitalist theory assumes that the profit motive will

inspire entrepreneurs to produce as efficiently as possible, entrepreneurs do not always do so. This may be so because the entrepreneur wants to advantage a relative or friend. It may also be so
because the management of large corporations, as personnel distinct from the shareholders, prefers the good will of the public to
maximum profit.[16] Or owners and managers of assured markets
may be uninventive or complacent, as British businessmen basked
in the twilight of Empire for the first half of this century. Conversely, owners and managers may concentrate on short-term
profit at the expense of long-term business growth.

Capitalist theorists may also overstate the willingness and desire of entrepreneurs to compete. Entrepreneurs may be more anxious to protect their share of the market than to increase it, and so
entrepreneurs may conspire with their competitors to divide the
market. Even without conspiracy, capitalism leads to monopolies
and oligopolies because successful competition necessarily elimi-
nates less efficient producers. Once monopolies or oligopolies are
established in the industries requiring large capitalization, new
competition may be effectively excluded from the market. Thus,
paradoxically, successful capitalist competition not subject to public regulation tends inevitably to destroy capitalist competition.

Indeed, many economic enterprises are natural monopolies
unamenable to competition. Utilities are prime examples of such
natural monopolies. There is little or no economic incentive for
two electric companies to operate in the same geographical area.
Once one electric company is operating in an area, no other
company will find it sufficiently profitable, or profitable at all, to
enter the market, although other modes of energy may provide
competition. Without public regulation, natural monopolies will
be free to charge consumers almost any price they choose.

Capitalist theory assumes that citizens should and will prefer
their long-term economic good to other social goals. But as Aquinas noted, economic good is subordinate to human good. Maximi-
zation of economic productivity may conflict with the social goals
proper to human development, and it is unreasonable to insist that
those social goals be subordinated to the economic good of the
community. The rank of economic good in the hierarchy of social
goods depends on society's allocation of values, and so society
itself does and should regulate the marketplace to protect noneconomic values.

Pure laissez-faire economics invites political instability. On the
one hand, professional and business classes are able to maintain

high standards of living and to consume conspicuously. On the other, the capitalist ethic encourages workers to expect higher standards of living as the reward for their labor. The resulting contrast between the conspicuous economic privileges of the few and the unfulfilled economic expectations of the many will threaten to undermine the legitimacy of the political regime sponsoring capitalism. Paradoxically if not inexplicably, the more economic growth laissez-faire economics delivers to developing nations, the more political instability it engenders, at least in the short run, by the unequal distribution of wealth which results.

Pure capitalist theory claims that any inequalities resulting from marketplace economics are justified on the basis of the marketplace's long-term benefits to society as a whole. But all citizens contributing to the good of the community, whether in economic or other ways, have a fundamental right to receive economic benefits sufficient for their human development. Therefore, no cumulus of the "greatest good of the greatest number" can justify excluding some citizens contributing to the good of the community according to their abilities, even temporarily, from minimal economic benefits. The pure capitalist theory would, in effect, treat some human persons as mere means to the economic benefit of others, and no end, however morally good, can justify the use of morally bad means.

Critics from Aristotle to Marx have argued more broadly that capitalism is intrinsically linked to greed, and the history of capitalism provides evidence to support the charge. According to critics, the acquisitive motive fundamental to capitalism debases the humanity of participants. Aristotle and Aquinas argued that entrepreneurs are debased because the quest for money displaces the quest for higher human goods. Marx argued additionally that workers are debased because capitalist society values workers only as commodities to be bought and sold in the marketplace. And religious leaders have complained about the materialism of capitalist society generally. The charge is very serious, and it is not easy to answer either in terms of capitalist theory or in terms of the historical record. Moderation of the acquisitive appetite may be possible in capitalist society—but not on pure capitalist principles.

At the opposite pole of the capitalist model is the communist model of economic distribution. Communism, rejecting capitalism both for its dehumanizing tendencies and the economic inequalities which result, would have each individual share in the burdens and benefits of economic production and consumption on a basis of complete equality. According to the classical Marxist principle, each should contribute to production according to his or her ability, and each should receive from the product according to his or her need. Critics may argue that communism, at least in Soviet practice, has failed to live up to the principle of equal distribution of economic benefits. An easy case can be made that government leaders, party officials, and production managers in the Soviet Union and the communist countries of Eastern Europe receive perquisites that separate their lifestyles from those of workers and farmers. The superiority of economic rewards to these leaders, officials, and managers may be less than the superiority of economic rewards their capitalist counterparts reap in the West, but the rewards are nonetheless superior. Such criticism fails to challenge the communist ideal itself. The question remains how well the communist ideal would satisfy principles of institutional and distributional justice.

However much communism may claim, at least in theory, to satisfy a goal of equal distribution, it fails utterly to satisfy the principle of freedom. Communist systems allow individuals little or no freedom to determine what they will produce or consume. Communist systems may seem to treat all citizens equally in the sense that distinctions between citizens in such matters as occupation and housing are made on the basis of objective factors, such as talent and community need. But leaders determine what citizens will produce and consume, and individual citizens are thereby denied the freedom of making those determinations for themselves. Moreover and more importantly, as Chapter 2 indicated, Marxist-Leninists deny personal freedom in the world of spirit as well. To achieve the goal of equal economic distribution, Marxist-Leninists identify the economic good of individuals with their human good and would ruthlessly reconstruct the consciousness of individuals who do not agree. As pure capitalism would effectively eliminate the common good, except as the sum total of individual economic goods, so pure communism along Marxist-Leninist lines would effectively eliminate individual goods, except as these are identified with the economic good of the community. Individuals under Marxist-Leninist regimes are simply parts of the social organism, and they are permitted only that freedom of self-realization that coincides with community realization.

Communist systems also pay a price for their ideal of equal distribution in the size of the product to be distributed. Production management in state-controlled economies has regularly proved

less efficient than production management in free-market economies, and workers and farmers in state-controlled economies seem
to work less energetically and efficiently for abstract communal
goals than workers and farmers in free-market economies do for
private gain. In the field of agriculture, the Soviet regime has
reluctantly recognized this fact for many years and granted farmers
some measure of private enterprise in order to increase production.
Communists argue that workers and farmers would act differently
if and when their social consciousness has been effectively reconstructed. Perhaps so, but more than six decades of attempts to
reconstruct social consciousness in the Soviet Union have failed to
do so. And in any case, it is difficult to see how and why reconstruction of social consciousness would make state-controlled managers of production any more efficient than they now are. If communist systems of equal economic distribution were so to affect the
size of the product as to leave all producers worse off than they
would be under a free-market system with some unequal distribution, the equality of the distribution would be of small comfort to
producers; the amount of the product to be distributed is as much a
part of the common good as the degree to which the product is
distributed equally.

Communist theorists argue that equal economic distributions
will be acceptable to workers, and that the unequal distributions
under capitalism will not. If workers act rationally in accord with
communist theory, they are destined to share equally in the benefits of the economic system over the long run and should be fully
content with their share, while workers in capitalist systems
should be ever more discontent with their unequal share in economic benefits. But workers in communist societies give no evidence of the satisfaction predicted by communist theory. In fact,
the Polish Solidarity movement and the persistent individualism
of Russian peasants give evidence to the contrary. There are several
reasons that may explain why this is so. First, most humans are not
willing to sacrifice other human goods to achieve purely economic
goods, that is, most humans do not agree that human good is
simply economic good. Second, in purely economic terms, workers
and farmers may perceive that equality of distribution is no substitute for more efficient production. Of course, workers and farmers
may also perceive that there is actually less equality of distribution
in communist systems than communist theory claims.

Communist theory can reasonably claim that it brings economic benefits to those who would be worse off in a pure capitalist
system. The unregulated marketplace will surely leave some citi-

zens badly off, and equal distribution will surely treat those citizens better. This is why communism is so appealing to those worse off in developing countries of the Third World. However, the communist promise of economic benefits to some citizens needs to be balanced against its economic and noneconomic costs: communism will leave most citizens economically worse off than they would be under capitalism over the long run because of the inefficiency of direct state control of production and consumption, and communism will leave all citizens worse off in terms of human freedom.

We can summarize conclusions about the relative justice of pure class, capitalist, and communist economic systems on two scales: (1) the freedom of all citizens to direct their lives as producers and consumers; (2) the ability of the systems to achieve distributional goals of productivity, of consensus on the fairness of distributions, and of minimum benefits to all citizens for their human development. A pure class system scores low on both scales. On the freedom scale, most citizens in a class system will have little or no freedom to direct their lives as producers and consumers. On the distributional scale, production will be static, disadvantaged citizens aware of inequalities will be dissatisfied with their share in the product, and only a few citizens will receive benefits sufficient for their human development. A pure capitalist system scores high on the scale of the freedom of individual citizens to determine for themselves what they will produce and consume without legal coercion. On the distributional scale, pure capitalism scores high on productivity but low on consensus for resulting distributions and on minimum benefits to those the marketplace leaves disadvantaged. A pure communist system scores low on the freedom scale. On the distributional scale, pure communism scores high on providing minimum benefits to all citizens but low on productivity and consensus. Thus, all three pure systems fall short of one or more elements of institutional and distributional justice. What remains to be considered is the degree to which a mixture of capitalist and communist principles might come closer to realizing substantial economic justice.

The Justice of Welfare Capitalism

Modern welfare capitalism contains both capitalist and communist principles of economic distribution: the free market allocates economic benefits initially, but the democratic regime partially redistributes marketplace results in favor of those less or not at all rewarded there. Welfare or democratic capitalism thus operates on two levels. On the first level, the economic level, the marketplace distributes the product according to the capitalist principle of merit. On the second level, the political level, the community partially redistributes the product according to the communist principle of need. Welfare capitalism attempts to retain both the capitalist concern for freedom and productivity and the communist concern for equal distribution. Unlike pure communism, welfare capitalism is willing to trade off some equality of distribution to achieve greater productivity. But unlike pure capitalism, welfare capitalism is willing to trade off some productivity to achieve greater equality of distribution.

Welfare capitalism scores high on the scale of the freedom of individual citizens to determine for themselves what they will produce and consume. While regulation of the market and redistribution of market results, of course, lessen the freedom of entrepreneurs, such regulation and redistribution increase the freedom of those whom the market disadvantages at least to determine what they will consume. In principle, welfare capitalism also stimulates productivity through the operation of the marketplace, makes the unequal results of the marketplace more socially acceptable by partial redistribution, and provides a "safety net" for those whom the marketplace leaves with insufficient economic benefits to support human life decently.

The actual redistributive effect of welfare capitalism in the United States and Western Europe should not be exaggerated. Entrepreneurs have managed to do quite well for themselves in the welfare state and to receive economic rewards far superior to those of workers. Entrepreneurs enter the economic and political arenas with overwhelming advantages of education, social position, and especially money. But a resolute majority can and does partially redistribute market results from those who have more to those who have less. This is at least the theory and minimal practice of welfare capitalist political economy.

Despite welfare capitalism's claim to moderate capitalism so as to make capitalism more egalitarian and socially acceptable, the democratic process alone will not guarantee that welfare capitalism will produce a relatively just society. Citizens in a welfare capitalist society may desire to acquire and consume material goods as immoderately, that is, without respect to proper human development, as citizens in a pure capitalist society. For one thing, partial redistribution of market results does not alter the way in which the marketplace produces its results in the first place. For another, citizens in a welfare capitalist society may bring the entrepreneurial spirit to the ballot box. If voters regard exercise of the franchise as simply an expenditure of capital on which they expect a return, then the acquisitive spirit of capitalism will dominate the political as well as the economic arena to the detriment of the common good. Welfare capitalism thus makes moral demands on citizens to moderate their acquisitive appetites at both the economic and political levels. A resolute democratic majority can restrain immoderate entrepreneurs. The more fundamental question about welfare capitalism is whether or not citizens of a democracy can restrain their own acquisitive appetites so as not to endanger long-term productivity.

As statesmen in democratic societies, fearful of voters' displeasure, may refrain from increasing taxes or reducing entitlement benefits when such policies are necessary for the common good, so citizens in democratic societies may use their political power to gain a larger share of consumptive goods at a heavy cost to long-term economic productivity. Peacetime rent control is a classic case of bad democratic economics. In a free market, landlords have an economic incentive to raise rents whenever demands for rental housing exceeds available supply. According to capitalist theory, the marketplace law of supply and demand will operate over the long run to increase rental housing supply to meet renter demand. But tenants facing large rent hikes in areas with rental housing shortages demand that their local governments control rents, and local governments often yield to the demands.

The long-term economic results of rent controls are as predictable as they are counterproductive of more rental housing. Landlords subject to rent controls may attempt to convert existing rental housing into cooperatives or condominiums. Tenants unable or unwilling to buy their apartments demand that local governments prohibit or limit conversions of rental housing, and local governments may do so. Landlords unable either to raise rents to the market level or to convert their buildings may then sharply restrict expenditures on their property, and property values will decline. Most important of all, from the viewpoint of future rental housing supply, rent controls on existing housing will discourage investments in new rental housing.[17] The end product of rent controls is that everyone affected by them is likely to suffer in the long run: landlords, future tenants, and even present tenants. Thus, apparently good short-term democratic politics can mean bad long-term democratic economics.

While a properly conceived public policy to deal with short-ages of affordable rental housing would proportion subsidies to the inadequate purchasing power of tenants or build public hous-ing for such tenants, rent controls indiscriminately subsidize all tenants. Conversely, while a properly conceived public policy to assist tenants with low purchasing power would pass the costs of subsidies on to the entire taxpaying public, rent controls assign the costs only to landlords. Landlords, of course, should share the costs of subsidies to needy tenants through income and property taxes, but so should others, including those tenants who are able to do so. Partial redistribution of market results to the needy from all those better off, however, has far less political appeal than a direct control of the marketplace that benefits all tenants indiscrimi-nately and burdens only landlords.

The ability of welfare capitalism to achieve a relatively just redistribution of marketplace results to those whom the market-place leaves less well off will depend on two essential qualities of politically responsible citizens: the intelligence to recognize long-term economic good and the virtue to moderate counterproductive acquisitive appetites. In the absence of that intelligence and virtue, citizens will blame their political leaders, not themselves, if limited resources fail to satisfy unlimited appetites. The challenge to dem-ocratic leaders today in the matter of economic distributions is to persuade the public to control its appetites in the interest of long-term economic productivity, and the challenge to democratic citi-zens is to recognize the wisdom of such a policy.

Control of majority acquisitive appetites is important not only to safeguard long-term economic productivity; satisfying a majori-ty's immoderate acquisitive appetites may also cost those, whom the marketplace leaves worst off, the minimum economic benefits for them to live decent human lives. Politically dominant working and middle classes, though willing to force upper classes to share economic benefits with themselves, may be unwilling to share their own economic benefits with the lowest classes. Limitations on natural resources suggest that working and middle classes in Western societies will have to make substantial sacrifices if the most disadvantaged in the marketplace are to receive economic benefits sufficient for their human development.

At a minimum level, access to work and a living wage are not at all matters of trading off equality to enhance productivity; access to work and a living wage are irreducible requirements of eco-nomic justice. No economic system will satisfy distributive justice if a minority of citizens, or indeed any citizen, is denied access to work and a living wage in the interest of majority prosperity. The greater economic good of most citizens cannot morally justify excluding some others from that share of material goods necessary to develop themselves as human persons. How to provide access to work, what constitutes a minimum wage for workers, and what benefits those unable to work or unable to find work should receive, of course, are questions that citizens of a democratic society collectively decide by exercising practical reason. But justice requires in this respect that any distribution of economic benefits to the most underprivileged should be determined neither by the marketplace nor by the greatest economic good of the greatest number.

Capitalism necessarily involves periods of economic expansion and contraction according to the law of supply and demand. Capitalism, without state intervention and management, cannot control calamitous cyclical fluctuations, as the Great Depression amply demonstrated. As a consequence, welfare capitalist governments should and do manage the marketplace to prevent runaway depressions and inflations. Second, capitalism necessarily involves the unemployment of workers in obsolescent and inefficient industries. In the heyday of industrial capitalism, the jobs created in new industries matched and even exceeded the jobs eliminated in older industries. Present-day automation of industrial and clerical tasks, however, costs far more jobs than it creates. Third, the new jobs generated by automation today require special skills that the most underprivileged do not have and cannot easily acquire, whereas the jobs generated by new industries before recent decades required no such skills.

In this context of cyclical unemployment and unemployment resulting from automation, government action is required to guarantee work for all. The government may do this indirectly by monetary and fiscal policies that stimulate the private sector to expand or directly by public works programs. Since the Great Depression, governments of Western capitalist societies have assumed a broad responsibility for the full employment of citizens. Government generation of employment, however, may come with a high price tag. Indirect stimulation of production may cause inflation, direct public employment will eventually mean higher taxes, and full employment will force employers to hire less efficient workers. Conservative economists and statesmen, therefore, argue that there has to be a trade-off between the level of unemployment and the costs of full employment.

Zero unemployment in a capitalist society is all but impossible. Even were the overall supply of jobs equal to the overall number of available workers, job supply and job demand would be unlikely to match one for one. Jobs may be available in California, but workers available in West Virginia. Or the jobs available may require skills that the workers available do not have. The government may help marginally to match workers with jobs by stimulating investment in areas of high unemployment and by job training programs, but much more government action is necessary to deal with the problem of structural unemployment. The central moral question about the government's response to the problem of structural unemployment is whether or under what conditions it is just to trade off the unemployment of some workers for the benefit of other workers and citizens.

First, it is well to remind ourselves what unemployment means for the human persons who are unemployed. In the absence of unemployment and welfare benefits, most of the unemployed over a long or short run will be deprived of the economic means necessary for human subsistence and development. Even with unemployment and welfare benefits, the unemployed suffer the psychological trauma of feeling useless and unwanted by society. Second, wages are the practical means whereby individuals have access to the property necessary for human development, and the human development of no individual citizen should be permanently sacrificed to the economic comfort of other citizens. Third, even if structural employment is a temporarily necessary means to bring inflation under control, the public has an obligation in justice to compensate those whom it causes to be unemployed with the economic means necessary for their human development.

There are those in modern society who are so disadvantaged that they are minimally or not at all employable. The disadvantages may be physical, as in the case of the disabled, or psychological, as in the cases of the mentally ill and the mentally retarded. The culture of poverty may also leave some without the necessary skills or motivation to secure permanent jobs. Except perhaps for providing job training, organized society can do little to eliminate these disadvantages. What organized society can and should do is to provide the material means for citizens so disadvantaged to develop their potential as human persons. In a capitalist culture, it is easy to forget that the disadvantaged contribute to the good of the community, although not necessarily to the economic good of the community. The disadvantaged, especially by the spirit of courage with which they bear their misfortune, contribute to the common human goal of living virtuously and so should receive from the community as a matter of right the material means to develop their potential as human persons.

Justice is a matter not only of society distributing wealth so that all members of the community may develop their human potential; it is also a matter of society rightly ordering conditions of work so that workers have the opportunity to assume responsibility for, and to perfect themselves by, their work. Marx and Engels accurately pointed out in their *Communist Manifesto* that, with the division of labor and the use of machinery in modern industrial production, work lost its individual character and its satisfaction for the worker.[18] If workers are not to perceive themselves simply as cogs in the machinery of the modern factory system, they need to participate in decisions affecting the work they do.[19] This is most important at the task level, that is, it is most important that workers have a say about the organization of their own labor, with worker participation extending to general industrial policy making. The means to implement this objective of greater worker responsibility for their work could take many different forms. One possibility is to make workers shareholders in the business they work for. A second is to encourage worker-run cooperatives. A third is to place worker representatives on company boards. Marx and Engels notwithstanding, there is no one simple solution to the problem of worker participation in decisions affecting their work.

In summary, welfare capitalism offers the greatest possibility of maximizing human freedom and achieving the distributive goals of productivity, social acceptability, and minimum benefits to those left least well-off in the marketplace. But welfare capitalism risks accommodating the principle of distributional economic equality both too much and too little; working- and middle-class majorities may demand too much equality with the rich to satisfy long-term economic productivity and yield too little equality to the underprivileged and disadvantaged to satisfy the latter's right to human development. In short, the ultimate justice of welfare capitalism depends on the virtue of moderation. This but echoes the thoughts of Aristotle (NE 5. 1; cf. Plato *Republic* 1. 2) and Aquinas (ST 2-2. 58. 5, 6) that justice in the most general sense is identical with moral virtue itself insofar as moral virtue is ordered to the good of the community.

The Justice of the International Economic System

We have been concerned in the preceding sections of this chapter with the justice of economic institutions and distributions within nations, especially developed nations. But there are other, perhaps more serious questions about the economic relations between developed and underdeveloped nations and the economic results for peoples of underdeveloped nations. The gulf between levels of production and consumption in developed nations and those in underdeveloped nations is tremendous. Three broad and interrelated questions about the justice of the international economic system, therefore, arise: (1) what obligations do developed nations owe to underdeveloped nations as a matter of justice? (2) to what extent do the current economic policies of developed capitalist nations contribute to poverty in underdeveloped but developing nations? (3) what policies should developed nations pursue to assist the economic development of underdeveloped nations?

To repeat an observation made at the beginning of this chapter, obligations of justice are distinct from those of charity. All humans belong to the same family, and the material resources of the world are designed for the common use of that family. Although the human family is not organized into one world society, citizens of affluent nations have obligations as a matter of justice to assist citizens of poorer, underdeveloped nations to lead properly human lives. Hobbes would have us believe that no order of justice exists unless individuals covenant to organize themselves into a single political unit under a sovereign power and a rule of law, but right reason recognizes a natural order of justice among as well as within nationally organized societies. Humans, wherever they live and however they are socially organized, share a common bond and enjoy a common right to use the world's natural resources for their proper development.

We repeat here another point stated at the beginning of this chapter, namely, that poverty is not necessarily the product of injustice. Underdeveloped peoples may be poor because they lack natural resources and so are vulnerable to natural catastrophes; lack of water and arable land in some parts of the world induce periodic or endemic drought and famine. Underdeveloped peoples may be poor because they are not organized to develop the natural resources they have; the traditional societies in many parts of the world are ill-disposed toward economic development. Underdeveloped peoples may be poor because they have high rates of population growth;[20] population increases in many parts of the world may absorb so much of the goods produced that little or none is left to invest in production growth. Developed nations may bear no responsibility for the existence of such poverty, but they do have responsibilities to assist underdeveloped peoples in times of disaster, to help them conserve and develop their natural resources, and to encourage the political modernization and family planning conducive to their economic development.[21]

There is a capitalist model of economic justice between developed and underdeveloped nations that mirrors in key particulars the pure capitalist model of economic justice within developed nations. The long-term solution to the problem of poverty in underdeveloped nations, according to this model, lies in the influx of public and private capital from developed nations. Public grants and low-interest loans from developed nations provide capital for an infrastructure on which private enterprise can build to develop the economies of underdeveloped nations. These public funds support irrigation projects to facilitate large-scale agricultural production, transportation systems to facilitate the movement of raw materials to factories and of products to market, and power plants to provide electricity for industrial production. Private loans from the banks of developed nations provide capital for local entrepreneurs to operate factories and farms. To this influx of capital from developed nations is added access to advanced technologies of agricultural and industrial production. Indigenous free enterprise, funded by capital from developed nations, will dramatically increase production, create jobs, and bring higher standards of living to the peoples of underdeveloped nations.

But the results of the capitalist model of economic justice, at least in the short run, have not been so rosy. Capitalist development has created new industrial jobs, but it has also eliminated many others; the machines which create jobs for mechanics and operators displace manual laborers. Capitalist development has stimulated the production of luxury items for affluent locals and export abroad, but it has decreased the production of more necessary items for nonaffluent locals. Worst of all from the viewpoint of the development goal itself, the capitalist model often results in a new outflow of capital from underdeveloped to developed nations.

Underdeveloped nations have interest and principal payments to meet on loans from the governments and banks of developed nations, and the interest rates on private loans have been high for over a decade. If markets for the exports of underdeveloped nations, especially the volatile markets for commodities, are weak, or if economic growth is slower than anticipated, insufficient capital will be generated to pay the interest and principal due on the loans. Moreover, multinational corporations expatriate the profits earned by their subsidiaries in underdeveloped nations to developed nations. In these circumstances, unless the governments and banks of developed nations make sizeable concessions, underdeveloped nations will be caught between a rock and a hard place. If they default, they cannot expect to receive any new loans, and if they adhere to the schedule of loan payments, they will have to squeeze capital out of resources their own citizens would otherwise utilize or consume.

The capitalist model of economic development has also had unfavorable results in the agricultural sector for the poorer farmers and urban masses of underdeveloped nations. Capital investment left to its own devices naturally flows into the production of those crops which bring the highest cash return. Concretely, this means that more crops will be produced for export and less for domestic consumption. If underdeveloped nations had the capital to import staples, and urban masses there the resources to pay for them, an export-oriented agricultural economy might have little or no consequence for domestic consumption. But most underdeveloped nations desperately need to conserve capital, and urban masses without state subsidies cannot afford to pay the price of imported staples. Similarly, the capitalization of agriculture and the introduction of modern farming techniques have had mixed results. The agricultural product has been increased, sometimes very dramatically. But more affluent and mechanized farmers have in the process eliminated poorer and nonmechanized farmers, and the latter, once landless, have drifted to the cities to swell the numbers of the urban poor.

Capitalization and technological innovation in farming risk ecological damage to the often limited land and water resources of underdeveloped nations. Capital investment stimulates farmers to overgraze their land and overconsume water, and technological innovation makes it possible for them to do so. Poorer farmers overwork their marginal land, leaving it vulnerable to wind and water erosion. Even the use of fertilizers and pesticides may reduce soil quality and pollute water supplies. All of these ecological problems exist in developed nations too, but the consequences of ecologically bad land use may be more serious for underdeveloped than for developed nations, and political elites in the developed nations are likely to be more sensitive to ecological consequences than political elites in underdeveloped nations.

From this line of analysis, it is not difficult to conclude that the capitalist model of economic development for underdeveloped nations, like the pure capitalist model for developed nations, fails to satisfy distributive justice either between developed and underdeveloped nations or between affluent and poor citizens within underdeveloped nations. Capital investment from the governments and banks of developed nations does stimulate production, but that growth principally benefits the investors of developed nations and the elites of underdeveloped nations. The capitalist model of economic development, as currently practiced and at least in the short run, seems to leave small farmers, farm workers, and the urban poor of underdeveloped nations worse off than they were before the process of development began. This distribution is not likely to be acceptable to most citizens of underdeveloped nations, and it certainly leaves many without the material resources minimally necessary for their human development.

At the other extreme from the capitalist model of economic development is the structuralist or neo-Marxist model. According to the latter model, the capitalist political, economic, and social structures of developed and underdeveloped nations necessarily produce economic maldistributions between and within both. The solution is to replace those capitalist structures with socialist structures. Structuralists are far clearer about what they reject than they are about what they propose. The most likely socialist system would be one along Marxist-Leninist lines; this at least has been the historical socialist system.[22]

We have already indicated why a pure communist system fails to satisfy principles of institutional and distributional justice, and we shall summarily repeat the reasons here. First, because a pure communist system equates economic good with human good, Marxist-Leninist regimes deny to citizens personal freedom in the world of spirit. Second, a pure communist system allows individuals no freedom to decide for themselves what they will produce and consume. Third, a pure communist system does not encourage production or efficiency, a matter of no little importance for underdeveloped nations with high rates of population increase. Lastly, precisely because a pure communist system denies human freedom and discourages production, it is not likely to win general acceptance from citizens over the long run. From an exclusively economic perspective, even if a communist system could avoid reduction of spirit to matter and its Marxist-Leninist consequences for human freedom, it would still discourage production and so ultimately leave all citizens in underdeveloped nations badly off or at least worse off than they would be with a free-market system.

Structuralists rightly direct our attention to the effects of the capitalist model, at least in the short run, for the human development of citizens of underdeveloped nations, but their pure communist alternative, especially one on strict Marxist-Leninist lines, would have equally bad or worse consequences for human development. Justice requires only that the capitalist model be restructured, not that capitalism be dismantled. From a practical perspective, for reasons we shall explain below, there are weighty obstacles to restructuring the international capitalist system, but a restructured capitalism is the only choice with a chance to achieve substantial justice.

As welfare capitalist regimes of developed nations can redistribute marketplace results in favor of those disadvantaged by the marketplace, so developed nations can redistribute the marketplace results of the international economic system in favor of underdeveloped nations disadvantaged by the international marketplace. Private capital from developed nations will not invest in the low-profit, labor-intensive modes of production that are conducive to the human development of peoples of underdeveloped nations. This means that the capital from developed nations for such modes of production must come from their public sector. The governments of developed nations could themselves fund agricultural cooperatives, for example, or subsidize loans from private capital for that purpose. The result might be less aggregate economic growth but more widespread human growth for the peoples of underdeveloped nations. With respect to those underdeveloped nations presently burdened by heavy debt, the governments of developed nations could assume the role of creditor from private banks and restructure the loans to make interest rates and repayment schedules more favorable to debtor nations. Lastly, the governments of developed nations need to regulate those activities of their own nationals destructive of the economic and political good of underdeveloped peoples (e.g., bribery of officials, cartels, promotion of harmful products) so that the corporations of developed nations would be held to the same standards abroad as they are at home.

The governments of developed nations to date have devoted only a very low percentage of their gross national products (1 to 2%) to assist underdeveloped nations. The failure of governments of developed nations to do more can be attributed in considerable part to the unwillingness of middle-class electorates to sacrifice a measure of their own level of consumption. Moreover, developed nations' corporations will resist efforts to regulate their activities abroad, and developed nations' bankers will oppose restructuring of existing loans and consequent loss of prospective profits. As with welfare capitalism within developed nations, the political task of forging electoral support for reforming the capitalist international economic system in favor of the disadvantaged peoples of underdeveloped nations will be formidable. In addition to moral arguments, however, the self-interest of developed nations in preventing Marxist-Leninist penetration of the Third World should counsel reconstruction of the existing international economic system with a view to redistributing marketplace results in favor of underdeveloped peoples.

But the capacity of developed Western nations to promote economic justice in many underdeveloped nations is severely limited by the traditional structures of the latter. In Latin America, for example, landowners, army, and Church have traditionally constituted the ruling classes. With the introduction of economic development, entrepreneurs and bankers have, for the most part, displaced the Church in that coalition of political elites.[23] A triarchy of landowners, army, and entrepreneurs now weds a laissez-faire economy to authoritarian politics. Such regimes are not only unjust but also a prescription for disaster for all concerned. On the one hand, lower classes, left with little or no voice in what they produce or consume, will be dissatisfied with the dislocations and maldistributions of a purely marketplace economy and be without political power to modify the dislocations and maldistributions. On the other hand, political elites in underdeveloped nations, owing no accountability to lower classes, lack any short-term incentive to share economic and political power with the latter. Unless political elites in underdeveloped nations recognize their long-term interest in doing both, social revolutions seem inevitable, and Marxist-Leninist regimes the most likely result.

Developed democratic nations can do nothing directly to modernize the political structures of developing nations. They can, however, indirectly influence the behavior of local political elites through their control over grants and loans. Public grants and subsidies by developed nations can be conditioned on economic and political reforms, and private bankers can be similarly required to lend only to underdeveloped nations whose governments agree to those reforms. Concededly, there is no strong constituency within developed nations for exercising this kind of political direction over economic aid and loans to underdeveloped nations, and a strong constituency of bankers and corporations against it.[24] But political elites in developed no less than underdeveloped nations should recognize that reform of the economic and political structures of the latter is to their long-term interest and act accordingly.

Complicating still further efforts to achieve economic and political reforms in underdeveloped nations is the necessity for developed nations to coordinate their policies toward those nations. In a world of independently organized national states, there is no international organization with authority to regulate the economic relations between developed and underdeveloped nations, and each developed nation must give its consent to common policies. As a consequence, conditions one developed nation might impose on its own nationals and recipient nations, another developed nation might not. Unless the major developed nations can coordinate their policies, the elites of developing nations will continue to find capital available somewhere without any economic or political conditions, and entrepreneurs and bankers will face no restrictions on their activities in underdeveloped nations from their own governments. This, of course, is another good argument for world government.

A salient feature of the decentralized international system is the multinational corporation. A corporation with subsidiaries in other countries may evade control by its own government or any other. If country A tries to regulate the activities of a multinational corporation, the corporation may be able to avoid the regulations by acting through subsidiaries in country B. Thus, a multinational corporation, acting through subsidiaries, may escape regulation by developed nations, and co-opted political elites in underdeveloped nations may be all too willing not to regulate activities of subsidiaries of the parent multinational corporation.

In sum, the present international economic system is severely defective from the viewpoint of justice. The defects are the product of a symbiosis between political elites in underdeveloped nations and capitalist investors from developed nations. It will not be easy to work out an effective strategy to remedy the defects, but considering the defects of a communist alternative, especially a Marxist-Leninist alternative, citizens and statesmen of the free world have no other choice for a chance of achieving substantial justice.

Notes

1. Several points should be noted. First, Locke includes life and liberty as well as estate in the term "property." Second, Locke would permit the legislative power to take any part of an individual's material goods through taxation for the good of society (TG 9. 138–40). Individuals contracting to enter organized society thereby consent to majority decisions by the legislative power. Therefore, despite Locke's preference for a laissez-faire distribution of economic benefits, his theory would seem to accommodate even a communist distribution of economic benefits as long as citizens through their representatives consent to it.

2. C. B. Macpherson, *The Political Theory of Possessive Individualism: Hobbes to Locke* (Oxford: Clarendon Press, 1962).

3. C. B. Macpherson, *Democratic Theory* (Oxford: Clarendon Press, 1973), p. 38.

4. For a description of differences between primitive and market economic systems, see Eric R. Wolf, *Peasants* (Englewood Cliffs, N.J.: Prentice-Hall, 1966), pp. 3–4.

5. While the internal economics of primitive societies raises no questions of distributive justice for members of those societies, the subsistence of primitive societies within modern societies does. When primitive peoples are situated within modern societies, the land primitives occupy often becomes attractive to moderns for economic development. The common good of modern society may justify some modification of land uses by primitive peoples, but a dominant modern society would surely violate justice if it were totally to subordinate the human values of primitives to the developmental values of moderns.

6. Although organized society bears no moral responsibility for acts of God, it does bear a moral responsibility to see that citizens affected by these acts continue to have those material goods necessary for human development, as we argue more fully later in the chapter.

7. As organized national societies are responsible for the consequences of acts of God in the distribution of economic benefits to citizens adequate for their human development, so developed nations have obligations to assist undeveloped nations without adequate natural resources for their citizens to live properly human lives.

8. I follow here the excellent organization and exposition of Thomistic texts by Anthony Parel, "The Thomistic Theory of Property, Regime, and the Good Life," in Anthony Parel, ed., *Calgary Aquinas Studies* (Toronto: Pontifical Institute of Mediaeval Studies, 1978), pp. 77–104.

9. The translation is from Thomas Aquinas, *Commentary on the Nicomachean Ethics*, trans. C. I. Litzinger, 2 vols. (Chicago: Henry Regnery, 1964), 1:49.

10. Parel (p. 92) claims that Aquinas, inconsistently with the principle of equal access to sufficient material goods, condones slavery. To reach that conclusion, Parel relies on the English Dominican Fathers' translation of the sentence "servus est possessio quaedam" (ST 2-2. 61. 3) as "a slave is his master's chattel." But "servus" in the medieval context means "serf," not "slave," and "possessio quaedam" means "a kind of possession," not "his master's chattel." The Latin text thus lends no support to the claim that Aquinas condoned making some human persons mere utilities for the benefit of others. We shall conclude later in the chapter that serfdom, at least as a permanent institution, is unjust, but serfdom is something less than full-fledged slavery.

11. Aquinas argues against a broad role for commerce on the basis of its consequences for the regime. Dependence on commerce would undermine the regime's independence, increase foreign influence in internal affairs, undercut the

status of nobles and knights, and cause excessive urbanization. But the worst consequence of more than a modest amount of commerce would be to corrupt virtue because citizens would seek to acquire riches rather than virtue.

12. The translation is that of Parel, p. 103. There is, to my knowledge, no English translation of the entire *Commentary on the Politics.*

13. The other principal way medieval businessmen circumvented the Church's strictures against usury was to receive recompense for lost business opportunities. As previously indicated in the text, Aquinas approved that practice. So did the Church.

14. On the distinction between traditional and modern forms of market economy, see Chapter 3, n. 2.

15. There are mixed models of economic distribution other than welfare capitalism. The British system, for example, involves elements from the class model in addition to those from capitalist and communist models. But corporatism, which gives still greater weight to traditional classes than does the British system, is probably the most important mixed model of economic distribution after that of welfare capitalism. Whatever the merits or demerits of the corporate economic model, it relies on an authoritarian political system.

16. See Adolph A. Berle, *The Twentieth Century Capitalist Revolution* (New York: Harcourt, 1954).

17. Investors, of course, may shy away from rental property for reasons other than fear of rent controls. Rental properties involve managerial responsibilities and long-term commitment of capital that may not be attractive to investors. No claim is advanced here that rent controls or fear of their imposition alone accounts for the current shortage of rental housing in many areas of the country, nor even that rent controls are primarily responsible for that shortage. The only claim is that rent controls aggravate the situation.

18. Karl Marx and Friedrich Engels, *Communist Manifesto*, ed. Samuel H. Beer (Arlington Heights, Ill.: Harlan Davidson, 1955), p. 16.

19. For an industrial psychologist's analysis to the same effect, see Frederick Herzberg, *Work and the Nature of Man* (Cleveland: World, 1966).

20. The link between high rates of population growth and poverty can be exaggerated. High rates of population growth will cause widespread poverty if cultivable land is devoted to the production of commodities for export or domestic luxury consumption. But high rates of population growth may be compatible with sufficiency if cultivable land is devoted to the production of commodities for general domestic consumption. What constitutes "overpopulation" is, to a considerable extent, a funciton of land use.

21. Historically, however, declines in rates of population growth have followed rather than preceded economic development.

22. We are without any historical model of a society organized economically on socialist principles and politically on such democratic principles as freedoms of speech, press, and religion, competitive political parties, and free elections, and it is conceptually difficult to see how the two sets of principles could be melded. One doesn't have to be a card-carrying Hobbesian to recognize that most individuals act instinctively in proprietary ways, and so that reconstruction of social consciousness is a necessary condition for a socialistically organized society. To reconstruct social consciousness along communal lines would accordingly seem to require extensive political control. Many entertain the hope that the Sandinista regime in Nicaragua will prove the possibility of organizing a society true to both socialist and democratic principles. The jury is still out on the future of democracy in Nicaragua, but the evidence available to date is not very encouraging.

23. When the economies of Latin American nations were organized on traditional, quasi-feudal lines, the Church there largely acquiesced in the economic

policies of political elites. With the modernization of Latin American economies on capitalist lines, however, the Church has become a vocal critic of the economic policies of political elites. As a result, the Church no longer plays an active role in support of political elites.

24. In underdeveloped nations, moreover, political elites can be expected to denounce Western pressures for reform as a new form of imperialism.

6

The Justice of Conventional War

Hobbes held the position that all war and modes of warfare are simply amoral, neither moral nor immoral.[1] There is no natural order of moral goods, and reason would be incapable of taking its measure in any case. Rather, nature endows individual humans and individual nations with rights to pursue their self-interests as far as they desire and have the power to achieve. Individual humans and individual nations should desire to live in peace, but they are not morally obliged so to live where there is no sovereign power to maintain peace. In fact, in this "state of nature," individuals and nations should prepare for war and even strike preemptively whenever expedient. A sovereign power, thanks to the covenant, exists in national societies to rescue humans from the warlike consequences of their naturally unlimited appetites. No such sovereign power exists in the international arena, however, and wars are thus inevitable. Nations are subject to no strictly moral norms either in their decision to wage war or in their conduct of warfare.

According to this model, the function of political reason in international relations is to appraise realistically the sources and limitations of national power and to devise aggressive or defensive strategies accordingly.[2] Hobbes would by no means recommend that nations resort lightly to war, or that they conduct war indiscriminately. But norms of war and warfare are those of national self-interest, not those of international morality. A decision to wage war will be "rational" only if the war is likely to be won, and if the likely gains from the war outweigh its likely costs. Indiscriminate war conduct will lead the enemy to retaliate in kind and so increase the costs of war. Thus, war and warfare may be irrational even though they can never be immoral.[3]

The Hobbesian position flies in the face of an "ethical fact," namely, that almost all people recognize that war needs to be morally justified. Because war involves killing, most people easily conclude that unjustified war and unjustified killing in war are immoral. Such was the overwhelming consensus about the moral guilt of the Nazi leaders responsible for World War II. It was on that basis, a crime against the international community, that 22 top Nazi leaders were tried at Nuremberg, and half of them executed.

Pacifism

Pacifists take a position diametrically opposed to that of Hobbes: all wars and modes of warfare are immoral. The Christian version of pacifism traces its origins to the teaching and example of Christ Himself. When Jesus announced the Good News about the advent of the Kingdom of God (Mark 1:15), He insisted on a change of values, and love of neighbor was basic to the new set of values (Matt. 22:39; Luke 10:27). In the Sermon on the Mount, Jesus taught His followers: "You have heard that it was said 'An eye for an eye and a tooth for a tooth.' But I say to you, Do not resist one who is evil. But if any one strikes you on the right cheek, turn to him the other also" (Matt. 5:38–39). And Jesus rebuked Peter for cutting off the ear of the High Priest's servant at the time of His arrest in the Garden of Gethsemene: "Put your sword back into its place; for those who take the sword will perish by the sword" (Matt. 26:52; cf. John 18:11).

While the pacific and nonviolent thrust of Jesus' teaching is clear, its applicability to the morality of war is not. The Gospel texts show only that Jesus advocated nonviolence by individuals toward other individuals; it is not necessarily the case that He advocated nonresistance on the part of public authorities to attacks on the community. Second, it is not clear whether Jesus prescribed the practice of nonviolence as a moral duty for His followers, or simply counseled the practice as a moral ideal. Third, with respect to the Sermon on the Mount, Jesus used typically Hebraic (exaggerated) modes of expression to emphasize His points, and so it is necessary to interpret passages in the Sermon cautiously before ascribing a strictly literal meaning to His words. Elsewhere in the New Testament, Paul declared rulers to be servants of God when they execute His wrath on wrongdoers (Rom. 13:4) and thereby seems to have legitimated public authorities' use of force to protect organized society.

Literary evidence from the first three centuries of the Christian era, though not extensive, indicates that early Christians disapproved the use of force by any Christian against another human being for whatever reason.[4] Origen,[5] Justin Martyr,[6] and Clement of Alexandria[7] can be cited for the view that early Christians interpreted Jesus' teaching as prohibiting service in the Roman army. The idolatrous rites often associated with Roman military life were an additional and perhaps decisive reason why early Christians avoided miltary service.[8] But early Christians did sometimes serve in the army and apparently without Church sanctions against them. The fragmentary evidence we possess suggests that the early Church took a dim view of military service by Christians, but that the decision to serve or not to serve was left to the conscience of the individual Christian.

With the Edict of Milan in 313 A.D. and the subsequent conversion of the Emperor Constantine, the attitude of Christians toward war changed radically, and the Christian practice of pacifism waned. With the Reformation, however, Christian pacifism revived. Anabaptists, Quakers, Mennonites, and smaller sects like the Brethren and the Shakers took a categorical position against war, and the Quaker William Penn instituted his "Holy Experiment" in Pennsylvania in the late seventeenth century. In the nineteenth century, Leo Tolstoy renewed interest in Christian pacifism beyond the confines of sect.[9] In the twentieth century, Mahatma Gandhi and Martin Luther King invoked the principle of nonviolence in their campaigns for Indian independence and racial equality, respectively. And young people disenchanted with American participation in the Vietnamese war in the late 1960s were attracted to the Buddhist tradition of pacifism.

No morally sensitive person, especially one who professes to be a follower of Christ, can fail to be attracted to a position that so universally respects human life or fail to be inspired by the often heroic witness of individual pacifists to principle. There are, however, two principal difficulties with the pacifist position, one theoretical and the other practical. As a matter of theory, it is difficult to see why those guilty of attacking the community should be entitled to do so without being forcefully resisted, or, conversely, why it is morally evil for those in charge of the community's welfare to resist wrongdoers forcefully. The practical difficulty is that a public policy of nonresistance would mean subjugation of a pacific community to a bellicose community.

Some pacifists claim that nonresistance will ultimately convert aggressors and oppressors. But that is likely to happen only in the long run, if at all, and humans should be able to live in a properly human way even in the short run. Moreover, the successful campaigns of pacifists like Gandhi and King were conditioned by the fact that Gandhi and King operated in basically free societies. Gandhi was ultimately successful because he could appeal to the moral conscience of the British electorate over the heads of colonial administrators, and King was ultimately successful because he could appeal to the moral conscience of the national American electorate over the heads of regional Southern officials. Substitute Nazi Germany for Great Britain in 1945 or for the United States in 1960, respectively, and the outcomes would hardly have been the same.

The Just War Decision

Hobbes held all war and warfare to be amoral, and pacifists hold all war and warfare to be immoral. Between those two positions is that of just-war theorists.[10] Just-war theorists ask moral questions about the decision to wage war, about the objectives that justify a nation's resort to arms and the conditions under which those objectives do so (*jus ad bellum*). And just-war theorists ask moral questions about the conduct of war, about the justice of the means of waging war (*jus in bello*). Just-war theorists attempt to distinguish between wars which satisfy justice and those which do not and between methods of warfare which satisfy justice and those which do not. We shall here examine just-war theory in the context of conventional war and so lay the foundation for assessing the justice of nuclear war in the next chapter, and the justice of supporting or intervening in revolutionary wars in Chapter 8.

As long as the Christian religion itself was outlawed or ignored in the Roman Empire, Christian pacifism and avoidance of military service had no political consequences. When, however, Constantine became a convert to Christianity, and Christianity the official religion of the Empire, an entirely new situation arose.

Christian rulers were now responsible for the welfare of Roman society, and this political power needed a moral theory on war that could serve as a basis for statecraft. Augustine provided the embryo of such a theory.[11]

Augustine distinguished between the use of violence in one's own behalf and the use of violence in defense of the community. In the case of the former, Augustine argued that Christian love barred Christians from the use of force in personal self-defense even when unjustly attacked. In the case of the community, however, Augustine argued that Christian love itself actually required Christians to come to the aid of others. Coercive power is necessary to remedy the public injustices caused by sinful individuals or nations, and so "it is the injustice of the opposing side that lays on the wise man the duty of waging wars" against foreign evildoers.[12] Statesmen and soldiers not only may wage war to prevent or punish injustices by foreign evildoers, they even have a duty to do so. In carrying out this duty, however, Christian statesmen and soldiers remain subject to the Great Commandment to love their enemies.

The embryonic Augustinian statement, which is hardly a formal theory, was amplified by Aquinas. Aquinas (ST 2-2. 40. 1) listed three requirements for the justice of war: (1) legitimate, that is, constitutional authority should make the decision to wage war; (2) war should have a just cause; (3) war should be waged with a right intention. To these conditions, the sixteenth-century philosopher-theologians, Francisco de Vitoria[13] and Francisco Suárez,[14] added three more: (4) the destructive costs of waging war, especially the cost in human lives, should be proportionate to the injustice sought to be remedied by recourse to war; (5) all peaceful means to rectify the injustice should have been exhausted before recourse to war; (6) there should be a reasonable hope that the just cause will prevail in the war.

Let us examine Aquinas's second condition. Self-defense to repel an armed invasion of a nation's territory is the most universally accepted cause justifying recourse to war. Of course, what constitutes national territory may be precisely the subject matter in dispute. Armed attack on national forces in international sea and air space is similarly accepted as a cause justifying defensive military action, although what constitutes international space may also be a subject matter about which the parties disagree. Aggression may take forms other than armed invasion of a nation's territory or armed attack on national forces in international space. One nation may be supporting insurgency against the government of another nation, or one nation may be building up an offensive arsenal for an attack on another nation.

In all of the above situations, nations that resort to war claim to act in self-defense. But the rights of nations may be violated in ways that do not directly involve defense of national territory or forces. Nations may seek to reclaim territory long out of their control, to recover nationals or nationals' property seized by another nation, to secure payment of debts, to guarantee the movement of international commerce, or to enforce treaty obligations. Sixteenth-century just-war theorists justified offensive wars to vindicate such national rights not only as a matter of justice but also for the long-term good of the international community itself.

To say that a nation has just cause to wage war does not settle the question whether or not the destructiveness of a war, especially in terms of human life, is proportional to the asserted just cause. Purely defensive wars may satisfy that principle of proportionality, but the same cannot be said of wars to vindicate national rights. "National rights" has always been an illusive and perhaps delusive concept. The principle of proportionality may well have required nations to forego resort to war to vindicate national rights, at least property-related rights, even in the sixteenth century. There is no urgency to resolve disputes between nations involving only property, and most property disputes among nations are subsequently settled by negotiation. Moreover, major powers have considerable economic leverage with which to vindicate national rights, at least against minor powers, without resort to war. Resort to war to punish foreign nations for killing nationals is obviously pointless with respect to the victims and unnecessary to insure the lives of other nationals against future loss; economic retaliation may adequately discourage future deviant behavior and induce later indemnification of victims' families. Limited resort to military action to rescue imprisoned nationals, of course, might be proportional to the costs of war both because the human freedom of the hostages is at stake, and because economic sanctions may work too slowly to have timely effect.

In the wake of World War I's terrible destruction, most nations of the world signed the Kellogg-Briand Pact. That Pact renounced war as an instrument of national policy and so at least implicitly repudiated the international legitimacy of offensive wars to vindicate alleged national rights. In the wake of destruction from a Second World War, Pius XII spoke for a broad international consensus of statesmen and professionals when he unconditionally condemned war as a means to vindicate national rights.[15] Because of the deadly destructiveness of modern technologies of war and the risk of worldwide war inherent in alliance systems, he argued, there can no longer be any possibility of a reasonable proportion between the destructiveness of modern war and the vindication of national rights.

The line between offensive and defensive war is often blurred. This is especially the case when one nation alleges that offensive military action is designed to prevent a military attack on itself. But however ambiguous many cases may be, the distinction between offensive and defensive war is a morally useful one. As we can distinguish day from night while being uncertain where one begins and the other ends, so offensive war can be distinguished from defensive war in their polarities. The recent Falklands War illustrates the point. The Argentines claimed title to the territory of the Falklands, and the British, who also claimed legal title, had occupied the territory without physical challenge for 150 years. In this context, the Argentine resort to military action against the British forces in the Falklands can easily be identified as an offensive war to vindicate national claims to the territory. Whatever the merits of the Argentine legal claims, this is precisely the type of situation that Pius XII had in mind, and which contemporary international consensus condemns, and the war itself demonstrated why the potential of modern technologies for destruction is disproportionate to the vindication of such national rights.

If Pius XII was correct in that no modern offensive war to vindicate national rights by developed nations can be proportionate to the resulting destruction of human life and property, then only defensive wars by them may satisfy the principle of proportionality. To know whether or not a particular defensive war does so, the benefits of resistance must be balanced against its costs. In calculating the benefits and costs of resistance, several considerations may be indicated. First, the objects to be defended by war can vary quantitatively and qualitatively. Defense of the homeland, for example, is of much higher priority than defense of a distant possession or protectorate, and defense of the homeland is of much higher priority than defense of international sea or air rights. On the other hand, the objects to be defended by war include intangibles like the long-term consequences for the nation attacked and even those for the international community. Cession of the Sudetenland to Germany in 1938 at Munich whetted rather than sated German territorial ambitions. Second, the destructive costs of war include the costs to the enemy and neutrals. Lastly, once hostilities have commenced, war losses are likely to escalate beyond original estimates.

The conditions that just war be a last resort and that just war have a reasonable hope of success are implicit in the principle of proportionality. Unless nations resorting to war have exhausted all reasonable efforts to negotiate, and unless those nations have a reasonable hope of successfully defending their rights, the justice of the cause will hardly outweigh the destructive costs of war.

A few words need to be added about the nature of success in war. The purpose of a modern just war is self-defense. At first blush, therefore, many defensive wars by small nations against large ones would seem to lack any hope of success. That evaluation depends on whether or not success is to be identified with short-term military success. Belgium in 1914 and Poland in 1939 resisted German invasion, not because those small nations hoped to be able to stop the invaders, but because they hoped their witness to the values of freedom and justice would arouse the international community to come to their aid. Belgian and Polish resistance rallied Western democracies against the German aggressors, and Belgium ultimately regained her freedom thereby. Belgian and Polish resistance additionally achieved the limited goal of preserving their freedom for a short period of time.

Aquinas's first condition for the justice of war requires that constitutional authorities should make the decision to wage war. That condition was historically concerned with the decision to wage offensive wars to vindicate national rights and designed to counter the centrifugal tendency of feudal lords to initiate such wars. But the condition is also relevant to defensive wars and to modern democracies like the United States. American line officers should be permitted to initiate large-scale defensive counterattacks only on the authority of the President of the United States. Otherwise, line officers may interpret minor border clashes as large-scale attacks without a determination to that effect by the President and his advisers. Moreover, even the deployment of U.S. troops in hostile situations over an extended period of time by the President now requires authorization from Congress.[16]

Aquinas's third condition requires that just wars be waged with a right intention. That condition includes not only the intention to wage proportionate war for a just cause, but also the intention to use only just means of warfare and to limit peace terms to achieving the just cause of war. The latter point needs particular emphasis. The purpose of just war is to right wrong, not to pain or plunder the enemy. But history amply attests how easy it is for nations to begin wars with the legitimate objective of self-defense and to end wars with illegitimate objectives of revenge or aggrandizement. In this context, moreover, it should be clear that no policy of "unconditional surrender" is consistent with just-war theory; just war is necessarily conditioned on justice.

We close this section on the just war decision with some observations about the centrality of negotiation and compromise in just-war theory. Just-war theory focuses on whether or when an act provoking war justifies a decision to wage war. But just-war theory also requires that statesmen seek to avert the act provoking war. There may be middle ground between all or nothing that will prove minimally satisfactory to disputing nations. If so, nations need to explore that ground. It is, of course, true that it takes two to negotiate, and so negotiation may be impossible with a nation like Nazi Germany.

Not only does just-war theory require negotiation to avert the act provoking war, it also requires negotiation after the act provoking war justifies a war of self-defense. The just cause is self-defense, and the right intention is to repel the enemy's attack on national territory or forces. There is no place in just-war theory for national pride or revenge. The fact that a defending nation is justified in waging war against an attacking nation does not necessarily mean that the attacking nation, however unjustifiable its resort to war, has no legitimate grievances against the defending nation. Moreover, if the attacking nation meets stiffer resistance than expected, or if other nations exert pressure, the attacking nation may become disposed toward peace. Even when under attack, a defending nation remains morally obliged to negotiate peace and to settle as far as possible the grievance which provoked the attack. Regional and international organizations can facilitate peace negotiations in conflicts between minor powers, and Article 42 of the United Nations Charter authorizes the Security Council to intervene in local conflicts when necessary to maintain international peace.

The Just Conduct of War

Just-war doctrine is concerned not only with the justice of waging war (*jus ad bellum*), but also with the justice of the way in which war is waged (*jus in bello*). Defending nations are morally justified to wage war only to maintain or restore justice. Thus, in the conduct of war, they are morally obliged to distinguish between the guilty enemy and the innocent enemy in the conduct of war. To the extent that harming or killing attackers is necessary to maintain or restore justice, those who commit injustice against the defending community by attacking it forfeit their right not to be harmed or killed. But those who are not committing injustice against the attacked community retain their right not to be directly harmed or killed. And so nations, like individuals, are not morally permitted directly to harm or kill innocents.

This moral principle discriminating between the enemy guilty and the enemy innocent of injustice, elaborated by de Vitoria and Suárez, was translated into a principle of international law. From the sixteenth to the twentieth century, belligerents recognized a distinction between attacks on military combatants and attacks on civilian noncombatants. The laws of war permitted the former but prohibited the latter. That is to say, noncombatants were entitled to absolute immunity from direct attack. This made sense in the context of premodern warfare, but modern warfare is quite different. In modern warfare, the whole nation's productive capacity is geared toward the war effort, and civilian producers of military equipment are as much participants in the waging of modern war as soldiers. The distinction between military combatants and civilian noncombatants is thus blurred in modern war and can no longer be the basis for distinguishing between the guilty and innocent enemy.

But the distinction between those guilty and those innocent of injustice remains critically important. Some civilians may indeed be guilty of participating in the waging of modern war, but many other civilians are not. Even the broadest definition of those guilty of committing injustice against the attacked community would surely exclude the elderly, children, and the sick from the category of guilty enemy. It all depends on the contribution that civilians are making to the war effort. Michael Walzer puts the point neatly:

> The relevant distinction is not between those who work for the war effort and those who do not but between those who make what soldiers need to fight and those who make what they need to live, like the rest of us. When it is militarily necessary, workers in a tank factory can be attacked and killed, but not workers in a food processing plant. . . . An army, to be sure, has an enormous belly, and it must be fed if it is to fight. But it is not its belly but its arms that make it an army. Those men and women who supply its belly are doing nothing particularly warlike. Hence, their immunity from attack.[17]

But there are many circumstances, indeed typical circumstances, in which military objectives are unobtainable without the loss of innocent lives. The question, therefore, arises whether or when military action against military targets will justify the incidental killing of innocents. The principle of double effect is the applicable principle to answer that question. According to the principle of double effect, indirect killing of innocents is morally permissible in a just war where the killing is unintended, where the killing is not itself the means of destroying a military target or achieving a military result, and where the killing is proportionate to the importance of the military target.

Application of the principle of double effect to warfare excludes deliberate killing of innocent enemy civilians and so includes a principle of discrimination. Nations are not morally permitted deliberately to harm or kill innocent civilians simply to increase the cost of war for the enemy. Nor may nations deliberately harm or kill innocent enemy civilians in order to achieve a military or political objective. For example, a just belligerent may justly seek to demoralize the enemy's civilian population so that the enemy's leadership will be induced to discontinue the war. But deliberately to harm or kill innocent enemy civilians in order to achieve that objective would be unjust.

The area bombing of German cities during World War II is a classic case of warfare that is morally unacceptable in terms of the principle of discrimination. Area bombing consisted of block-by-block incendiary bombing of residential areas; no military or industrial targets were involved. In one view, earlier indiscriminate German bombing of English cities justified retaliatory indiscriminate Allied bombing of German cities. In another view, the collapse of German morale, with the consequence of shortening the war, justified the Allied residential fire-bombing. Sir Arthur ("Bomber") Harris, Commander in Chief of the R.A.F. Bomber Command, put the case in simpler terms: to bomb anything in Germany was better than to bomb nothing. But none of these explanations satisfies justice.

The American atomic bombing of the Japanese cities of Nagasaki and Hiroshima in 1945 similarly involved an intention to kill and maim innocent enemy civilians in order to shorten the war.[18] Then-Secretary of War Henry L. Stimson subsequently articulated the strategic reasoning that led him and other American statesmen to approve the atomic bombing of the two Japanese cities: the bombings would cause Japanese morale to collapse, induce Japanese leaders to surrender, and save one million Allied casualties which an otherwise necessary assault on the Japanese mainland would cost.[19] But the bombings killed 105,000 Japanese civilians and maimed 105,000 more. There can be no moral trade-off between potential or actual military casualties and innocent enemy civilian lives because innocents have done nothing to deserve death.

On just-war principles, then, just warriors should not directly target innocent civilian populations. But just-war theorists, also applying the principle of double effect, would morally justify the proportionate, incidental killing of innocent enemy civilians as the result of attacks on legitimate military targets. Indeed, it is inconceivable that war could ever be waged without some incidental killing of innocent enemy civilians. How many such incidental deaths are proportionate to a military objective, of course, is no easier to define than civilian guilt or innocence in the waging of modern war. But some military objectives are surely more important than others, and the time-factor in warfare may give a lesser military target more importance than it would otherwise have. The principle of proportionality does not solve the moral problem concerning the impact of warfare on the lives of civilians, but the principle does pose an essential moral question: does the conduct of war result in the unjustifiable destruction of innocent human life?

International convention has regulated some war conduct. For example, international convention prohibits the use of poison gas, and international convention requires humane treatment of prisoners of war. Belligerents, of course, have a strong practical reason for observing these conventions, that is, were one belligerent to violate them, the other would be likely to do the same. But belligerents also have moral duties to observe international conventions, at least as long as the enemy observes them; belligerents should cooperate to reduce, not increase, the destructiveness and hardships of war.

The war against Nazi Germany may well be the clearest case in all history of a war waged for a just cause. And yet Allied war conduct was sometimes unjust; the injustice of Allied fire bombing of German residential areas has already been cited. Unjust Allied war conduct was relatively untypical, however, and most commentators have been unwilling to condemn the justice of the Allied cause on that account. On the other hand, the justice of a belligerent's cause would be vitiated were the belligerent systematically to wage war in unjust ways. Critics of American war conduct in Vietnam allege that pervasive American violations of the principles of discrimination and proportionality effectively made that war a war on innocent civilians. Whether or not the critics are right in their interpretation of American war conduct in Vietnam, they rightly link war conduct to the overall justice of war. Pervasively unjust war conduct is inconsistent with the concept of just war.

Notes

1. In a similar vein, the Prussian strategist Karl von Clausewitz cynically described war as the continuation of foreign policy "by other means." Karl von Clausewitz, *On War*, ed. Anatol Rapoport (Harmondsworth, Eng.: Penguin, 1968), p. 119.

2. Hans J. Morgenthau was perhaps the most articulate contemporary spokesman for this view of international politics; see his *Politics among Nations: The Struggle for Power and Peace*, 5th ed. (New York: Knopf, 1972).

3. To their credit, Enlightenment philosophers after Hobbes championed development of international law and international organization on the basis of mutual self-interest. But the success of legal codes to regulate and organizational structures to control international conflict ultimately depends on the willingness of sovereign nations to cooperate to achieve long-term goals of the world community at the price of short-term national goals. As with national politics, so with international politics, nations will not do so unless reason apprehends cooperation as a moral good, and unless will restrains appetite accordingly. In the final analysis, amoral self-interest alone, however enlightened, will not guarantee the perpetual peace sought by Enlightenment philosophers and their heirs.

4. For a brief survey of patristic comments on war, see Alfred Vanderpol, *La doctrine scholastique du droit de guerre* (Paris: Pedone, 1919), pp. 16–23.

5. Origen *Contra Celsum* 3. 8.

6. Justin Martyr *Trypho* 110.

7. Clement of Alexandria *Protrepticus* 11. 116; *Paedagogus* 1. 12.

8. See Edward A. Ryan, S.J., "The Rejection of Military Service by the Early Christians," *Theological Studies* 13 (March 1952):1–32.

9. Leo Tolstoy, *On Civil Disobedience and Nonviolence* (New York: Mentor, 1967).

10. There are other middle positions on the morality of war between those of Hobbes and pacifists. One is that the will of God, not human reason, determines whether or not wars and methods of war are moral. This was the view of biblical Jews: those wars and methods of warfare that Yahweh sanctioned were just, and those wars and warfare that Yahweh did not sanction were unjust. On the one hand, Yahweh could sanction war against the enemies of Israel in order to achieve and preserve the promises of the Covenant (Jer. 21:5; Isa. 10:5–6; Isa. 63:10). On the other, Yahweh could sanction war against Israel in order to punish her for violations of the Covenant. In the Song of Moses (Exod. 15:1–18), Yahweh Himself is described as a warrior. Indiscriminate destruction of enemy populations is morally approved (Deut. 20:10–18). In the case of enemies within Israel's borders, the *herem* or sacred ban required destruction of everything associated with the enemy (Josh. 6, 7). In the case of enemies outside Israel's borders,

destruction of the male population and sale of women and children into slavery were required. Even in this context of "holy wars," the prophets Isaiah (2:4) and Micah (4:3) held out the vision of a world without war, and the book of Leviticus (19:18) prescribed love of neighbor.

Several general observations can be made about appeals to the divine will as a norm to justify war and warfare. The divine will does not manifest itself in all situations, and so it is not a comprehensive norm whereby humans can determine the morality of war and warfare. Second, human agents are adept at appropriating the will of God to their own. Third, events narrated in the Old Testament need to be subjected to careful exegesis before they can be attributed to divine rather than human will.

Another middle position is that of Aristotle. Aristotle and the ancients generally adopted a rather matter-of-fact attitude toward war. War was indeed an undesirable event but also one that was unavoidable. Aristotle's main concern was with the internal right ordering of Greek city-states, and he thought such right ordering would minimize armed conflicts among them. Aristotle was concerned to minimize armed conflicts rather than to articulate moral norms to govern them. Aristotle argued for a *polis* big enough to resist attack but not so big as to invite attack by envious neighbors (*Politics* 2. 7). This norm of moderation, in keeping with his general ethical norm, may not seem too dissimilar to that of Hobbes.

Aristotle, however, expressly approved only three objectives of military training (*Politics* 7. 14): (1) preservation of the *polis* from subjection by others; (2) vindication of the nondespotic leadership of the Athenian *polis* over the Greek *poleis* for their own benefit; (3) mastery over non-Greeks. The first objective of military training is clearly consistent with just-war theory. The second objective undoubtedly reflects Athenian feelings of superiority over other Greek city-states by reason of Athens' cultural achievements, but it does at least exclude military training for the purpose of achieving despotic rule over other Greek states. And the third objective of military training was based on the common Greek assumption that non-Greeks were by nature unfit to rule. (Aristotle linked war against non-Greeks to hunting, and Greeks were said to acquire property rights over non-Greeks by war [*Politics* 1. 8].) Stripped of its racist presupposition, that objective hardly seems consistent with a notion of justice.

A third middle position is that of those Protestant theologians who, relying on the Reformers' central principle that original sin taints the whole human enterprise, deny any possibility of a just war. The Protestant principle of man's irremediable sinfulness, with its attendant effects on reason, leads these theologians to conclude that all humanly mediated political efforts are immersed in sin and contaminated with injustice. War, like all politics, is incapable of moral legitimation. Reinhold Niebuhr forcefully articulated this theological view of national and international politics. See his *Christianity and Power Politics* (New York: Scribner's, 1940).

But because Niebuhr also acknowledged that war in some circumstances (e.g., self-defense) might be the lesser of two evils, Niebuhr can be said to have "justified" such wars, albeit not to legitimate them morally, and so to reach conclusions about the morality of war not very substantially different from those of just-war theorists. Niebuhr's position is thus distinguishable from that of pacifists; pacifists not only condemn all war as immoral, but they would never even limitedly justify war as the lesser of two evils.

11. Augustine *De civitate Dei* 19. 7, 12, 13, 15.

12. Augustine *De civitate Dei* 19. 7. The translation is from *The City of God*, ed. David Knowles (Harmondsworth, Eng.: Penguin, 1972), p. 862.

13. Francisco de Vitoria, *Relectiones: De Indis et de jure belli*, ed. Ernest Nys (Washington, D.C.: Carnegie Endowment for International Peace, 1917).

14. Francisco Suárez, *De triplici virtute theologica: De caritate, disp. 13* [*de bello*] in *Selections from Three Works*, ed. James Brown Scott, 2 vols. [Latin text and English translation] (Oxford: Clarendon Press, 1944).

15. Pius XII, "Già per la sesta volta," *Acta Apostolicae Sedis* 37 (1944):18; John XXIII, "Pacem in Terris," *Acta Apostolicae Sedis* 55 (1963):291, summarily affirms the same position.

16. Congress asserted the right to supervise presidential disposition of American armed forces in hostile situations in the War Powers Resolution, 50 U.S.C.A. #1541–48. The constitutionality of the resolution has not yet been judicially tested.

17. Michael Walzer, *Just and Unjust Wars* (Harmondsworth, Eng.: Penguin, 1980), p. 146.

18. Some critics of the Nagasaki and Hiroshima bombings contend that the bombings were immoral because of the nature of the bombs. I argue that, independently of the atomic dimension of the bombings, the bombings were immoral because of their targets.

19. Henry L. Stimson, "The Decision to Use the Atom Bomb," *Harper's Magazine* 94 (February 1947):100–101, 106–107. Some historians have challenged Secretary Stimson's factual assessments. But the fundamental moral judgment about the bombings from the perspective of just-war theory does not depend on the accuracy of Stimson's factual assessments, assessments understandably chancy in the stress of wartime decision making.

7

The Justice of Nuclear War

In this chapter, we shall explore the application of just-war criteria to nuclear war.[1] In U.S. strategy, the purpose of nuclear weapons is to deter, not wage, war. It is nevertheless necessary to consider the morality of their possible use.[2] Effective deterrence rests on a credible threat to wage a successful defense against an aggressor or to inflict unacceptable damage on him. A deterrent will in fact deter a potential aggressor only if the aggressor believes that the potential victim has the capability and will to fight back. Nuclear deterrence may fail, moreover, and if it did fail, the choice for the victim of aggression would be between some form of nuclear war and surrender.

Some would claim that it is impossible to apply just-war criteria to nuclear war because the Soviet Union does not accept those criteria. In this situation, they argue that the U.S. and the Soviet Union live in a Hobbesian universe where no moral rules beyond the desire to avoid war apply, and they see no moral problem about targeting Russian civilians as a means of national survival. But statesmen no more than private citizens are absolved of moral responsibility for their own actions simply because others act, or are prepared to act, immorally.

At the other end of the spectrum, some would claim that it is impossible to apply just-war criteria to nuclear war because nuclear war as such is morally repugnant. It may well be the case that all forms of nuclear war are unjust because of their consequences for humankind. But if so, this can only be discovered by applying just-war criteria to different forms of nuclear war. We shall examine broad categories of nuclear strategies and tactics one-by-one to assess whether or not they conform, or could conform, to just-war criteria.

Some just-war criteria, however, can be applied to nuclear war without considering different strategies and tactics. One just-war criterion requires that competent, that is, legally constituted, authority and only competent authority make the decision to wage nuclear war. The President of the United States, as Commander in Chief of U.S. forces, has the constitutional authority to resist an armed attack on the United States or a nation toward whose defense the United States has treaty obligations, although Congress has the technical constitutional authority to declare war. Critics worry that the effects of conventional or nuclear war would disrupt what defense specialists call C^3: the chain of command up to the President, control of those lower in the chain by those higher up, and communications between levels of command. Specifically, critics fear that local commanders would be left to make decisions about the use of nuclear weapons, or that they might use the weapons without authorization from superiors to do so.[3] Since we do not have access to Department of Defense contingency plans, we cannot well evaluate how serious the problem is, or how difficult solutions for the problem would be. The problem could indeed be serious, but there is no reason to think any existing C^3 weakness permanently irremediable.[4]

Both the United States and the Soviet Union continue to centralize and improve their C^3 and intelligence, although this is very expensive. (The U.S. plans to spend $20 billion in the next four years in this effort.)[5] Centralization will lessen the likelihood of accidental launch and the need to delegate authority to launch. Whether or not comprehensive centralization is feasible is another question. Above-ground radar stations are highly vulnerable, and many scientists claim that a few nuclear airblasts would set off an electromagnetic pulse that would destroy the electronic and communication networks of centralized systems.[6]

All Western nuclear strategies and tactics threaten to use nuclear weapons only to defend the West against Soviet military aggression, and the justice of such a cause should be clear. The stake for the West in resisting Soviet aggression is freedom itself, not merely economic freedom but also and especially freedoms of mind and heart. We have only to look at the Marxist-Leninist regimes in Eastern Europe to know what the consequences of successful Soviet aggression would be for freedom in Western Europe and the U.S.: no freedom of speech, no freedom of the press, little freedom of religion, and limited freedom in the arts and sciences. The justice of the Western cause of self-defense against Soviet aggression cannot be over-stressed; it is the cause of human freedom. To say that the Western cause in defending its people against Soviet aggression is just, of course, does not imply that all activities of Western societies are just. But on the question of defending human freedom, the cause of the West is unqualifiedly just and paramount.

As conventional just warriors should wage war with a right intention, so should nuclear warriors. Right intention means the intention to right wrong, not to inflict harm on the enemy, and to achieve a just peace. Right intention is difficult enough to maintain in modern conventional wars involving nation states. It is likely to be still more difficult during and after a nuclear exchange. One cannot, however, antecedently condemn all forms of nuclear war on the ground that they would lack right intention. At least one cannot do so as long as American policymakers maintain the position that the purpose of nuclear weapons is to deter aggression and to defend freedom as a last resort if deterrence were to fail, not to coerce any other Soviet action or inaction and certainly not to "win" a nuclear war.

Along with right intention goes a willingness to negotiate outstanding issues and to use all means to resolve disputes peacefully. This just-war criterion is not easy to apply when the potential enemy is ideologically committed to the overthrow of Western structures and values. On the one hand, all negotiations with ideologically pure Marxist-Leninists involve only temporary truces rather than a lasting peace. On the other, Marxist-Leninists have a mutual interest with the West in avoiding nuclear war, and the two sides may find acceptable areas of agreement on that basis. In any case, the West is morally obliged to avoid war by all means short of appeasement or surrender, although any hypothetical situation in which Soviet armed forces were attacking and overwhelming NATO conventional forces and/or one in which the Soviets initiated a nuclear attack on the West necessarily implies that peaceful means to resolve outstanding differences between East and West, if they have been tried, have already failed. The moral obligation to search for ways to end hostilities and resolve the conflict persists after a nuclear exchange. As with right intention generally, there is no a priori reason to exclude the possibility that the West might have negotiated in good faith before the Soviet attack, might have exhausted peaceful means before resort to nuclear defense, and would persist in seeking ways to end the hostilities after a nuclear exchange.

The Morality of Nuclear Warfare

The central moral issue in nuclear war is whether or not the destructiveness caused by nuclear weapons can ever be proportionate to the just cause of defending human freedom. Nuclear weapons would not only obliterate and maim life in the area of impact but also cause collateral damage to a much larger area, especially through radioactive fallout. Some fear that the explosion of many nuclear warheads could result in extinction of the entire human race. The just cause of defending human freedom would hardly warrant resort to a form of warfare that would destroy or maim most of the human race in whose name the warfare was waged. On the other hand, it is at least a priori possible that some nuclear strategy or tactic might satisfy the principle of proportionality. An examination of particular nuclear strategies and tactics may show whether or not any one of them does or might do so.

It is possible to imagine the discrete use of a nuclear weapon that would satisfy the just war-conduct principle of proportionality. For example, we might imagine the use of a nuclear weapon against an enemy tank formation in the countryside or a remote enemy airfield or an enemy missile base in the tundra. But the moral problem of proportionality in nuclear war is not simply to assess the balance between the destruction of human life wrought by use of a single nuclear weapon and the defensive value of destroying specific military assets of the enemy. The problem is rather to assess the balance between the destruction of human life wrought by the use of many nuclear weapons and the just cause of defending human freedom. Any use of nuclear weapons opens a Pandora's box of possible scenarios, the most benign of which would be a limited nuclear exchange, and the worst of which would be a nuclear shootout.

Any attempt to assess the proportionality of nuclear strategies and tactics to the just cause of self-defense labors under serious constraints. First, there are almost as many nuclear strategies as there are nuclear strategists. Second, most information about nuclear weaponry is classified and so inaccessible to laymen. Third, even with access to classified information, factual estimates about nuclear weaponry are only educated guesses. Fourth, the state of nuclear weapons technology is so fluid that, like generals fighting the last war, moralists may be passing judgment on yesterday's technology. Last, and perhaps most important, attempts to assess proportionality hinge in part on assessments of the likely behavior of American and Soviet policymakers in crisis situations. We have difficulty enough estimating the likely behavior of American policymakers, and we have still greater difficulty entering the mind-set of Soviet policymakers to assess their likely behavior.

Despite these serious constraints, we shall attempt, as best we can, to inquire into the morality of different uses of nuclear weapons in defense of Western freedom.

For purposes of moral analysis, we can identify five potential targets of nuclear strategies and tactics: (1) countercity targets, (2) countercontrol and counterintelligence targets, (3) heartland counterforce military targets, (4) theatre or intermediate counterforce military targets, (5) tactical or battlefield counterforce military targets. To these may be added strategies of defending populations against attacking enemy missiles (e.g., antiballistic missiles, civil defense) or of defending satellite and radar systems against antisatellite or antiradar devices. None of these strategies or tactics excludes the others, some strategies are usually combined with others, and all strategies are complemented by tactical considerations. Let us consider the five offensive nuclear strategies and tactics and the defensive strategy against attacking enemy missiles in turn.

Countercity Strategy

The first strategy, which targets Russian cities, was the main U.S. strategy even into the second decade of the atomic era. At the opening of the era, the United States had overwhelming superiority in nuclear weapons and threatened "massive retaliation" to deter the Soviet Union from conventional attack on Western Europe. When U.S. nuclear superiority declined toward parity with the Soviet Union, the United States continued to rely on a countercity strategy to deter a Soviet nuclear attack but conceded that the United States itself would suffer catastrophic damage in a countercity nuclear exchange. The strategy was then described as "mutual assured destruction," and the very appropriate acronym "MAD" was added to our vocabulary.

According to the MAD theory, neither the Soviet Union nor the United States could afford to attack the other with nuclear weapons because each nation could absorb a first strike by the other with enough nuclear power intact to be able to inflict catastrophic and, therefore, unacceptable damage on the initiator of nuclear war. The Soviet Union, facing this prospect, would be deterred not only from a first nuclear strike but also from an all-out conventional attack on Western Europe because such an attack could escalate from conventional warfare to tactical use of nuclear weapons to a strategic nuclear exchange. The U.S. began to move away from an exclusively countercity strategy in the early 1960s and publicly disavowed the strategy in 1983.[7] Despite the public disavowal, many believe that the current deterrent policy remains basically a countercity strategy, and many experts still prefer a countercity strategy to any other.

Since there has been no nuclear war or conventional Soviet attack on Western Europe in the forty years since the end of World War II, proponents argue that the countercity strategy has been a successful deterrent. Whether or not the absence of nuclear war or a conventional Soviet attack on Western Europe can be attributed solely or principally to MAD is debatable, but the most serious questions concern the practical and moral consequences of a countercity strategy if deterrence were to fail. In other words, would it make any practical or moral sense to use nuclear weapons against Soviet civilian populations in the expectation of retaliatory nuclear attacks on our own civilian populations? It would surely be suicidal, and so there are few, if any, voices in favor of a countercity nuclear strike to defend Western Europe or even to retaliate in kind to a limited Soviet nuclear attack. That throws considerable doubt on the credibility of countercity strategy as a deterrent.

Countercity nuclear strikes not only would run counter to a Hobbesian interest in self-preservation; they also—and not least— would run counter to moral perspectives on the value of human life. Countercity nuclear strikes, both in terms of casualties to the civilian populations of the belligerents and in terms of collateral damage to the rest of the human race, would surely violate the principle of proportionality. In any case, we have already argued that the deliberate targeting of enemy civilian populations would morally fail to distinguish the innocent enemy from the guilty enemy in the matter of taking human life.

Countercontrol Strategy

The U.S. over the last decade has moved toward a different nuclear strategy, which we shall call countercontrol strategy. This strategy developed for a number of reasons. As already indicated, a countercity strategy would be both suicidal and immoral. Second, a countercity strategy lacks credibility as a deterrent precisely because it would be suicidal to carry out. Third, there was concern that the Soviet leadership might not attach the same value to civilian populations that the West does.

The principal targets of a U.S. nuclear countercontrol strategy are the Soviet apparatus of military and political control and the information-gathering systems on which that apparatus depends. The theory is that the Soviet leadership attaches the highest value to the maintenance of political and military control over their own and Eastern European populations, together with preservation of their military and industrial base. The threat to destroy these would, according to countercontrol strategic theory, deter the Soviet leadership from initiating nuclear war and indirectly even a conventional war against Western Europe. And proponents argue that to carry out the threat, were deterrence to fail, would be neither suicidal nor immoral.

To carry out a countercontrol strategy would, on the face of it, conform to one of the two key moral principles at issue, namely, the principle of discrimination. The avowed targets of a countercontrol strategy are the nerve centers of the Soviet political and military machinery, not civilian populations. Since the main Soviet apparatus of political and military control is in or near populated areas, however, the distinction of countercontrol from countercity strategy may be meaningless. In spite of the attempts to distinguish between the innocent enemy and the guilty enemy, the massive cost in human lives on both sides and the collateral damage to civilian populations of belligerents and neutrals would seem almost as disproportionate to even the just cause of defending freedom as in the countercity strategy. Moreover, Soviet missile systems could be put on some type of automatic pilot to cover the situation where Soviet control centers were destroyed or communication with them lost, and so direct casualties and collateral damage might be multiplied beyond control-center areas.[8]

Countercontrol strategists may well be right in supposing that Soviet leaders value their military and political control centers more than they do the lives of their citizens. But if so, it would seem to follow that the Soviets would be willing to defend their control centers at greater cost. U.S. weapon systems aimed at Soviet control centers threaten the Soviet system itself. To the Soviet leadership, a countercontrol policy may well appear to be an attempt to defeat them, not to deter them. U.S. efforts to shorten the delivery time for nuclear weapons to reach Soviet control centers and to develop the capacity to knock out Soviet satellites in space might provoke Soviet leadership to strike first before it lost control.

Counterforce Strategy

Another nuclear strategy is called counterforce. The targets of a counterforce strategy, if deterrence were to fail, would be Soviet missiles, airfields, military bases, troop concentrations, and so forth. Advocates of this strategy contend that it offers a realistic war-fighting posture if deterrence were to fail and, precisely for that reason, has greater credibility to deter a Soviet nuclear first strike. They envision a limited nuclear exchange as a last-ditch attempt to avoid surrender if deterrence were to fail. And they contend that counterforce strategy conforms to the moral principles of discrimination and proportionality.

It is generally conceded that counterforce strategy satisfies the principle of discrimination. The targets are clearly military, not civilian. But proportionality remains a problem both because no one can be sure how much death and destruction a strictly observed counterforce strategy would entail, and because no one can be sure whether or not a counterforce strategy could, in fact, be confined to the limits its advocates propose. Moreover, a U.S. counterforce strategy could destabilize whatever balance there is currently between the U.S. and Soviet nuclear arsenals, and this might provoke the Soviet leadership to strike first or tempt an American President to do the same.

There is uncertainty about the accuracy of counterforce missiles. Technologists describe the accuracy of weapons in terms of the area around the target within which 50% of explosives will detonate. Unless we understand exactly what that means, we might derive a false sense of comfort from an expert's assurance that a given nuclear missile has a circular error probability (CEP) of, say, a 400 yard radius from the target. No doubt ballistic experts would consider a 400 yard radius a highly accurate CEP for a missile that traveled thousands of miles before reaching the target, but a statesman or moralist should be apprehensive. The factual question to be asked is: where would the other 50% of the nuclear explosives detonate? The uninitiated might expect them to detonate within an arithmetically proportional additional area (say, an additional 400 yard radius). But, in fact, 50% of the nuclear explosives of the postulated missile could detonate *anywhere* outside the given margin of error.[9] This means that an errant missile could easily hit population centers and so inflict massive death and destruction.

Objection to such targeting on grounds of disproportionality might be as valid as that against countercontrol strategy and, a fortiori, against countercity strategy. On the other hand, not to target some important enemy military assets, especially nuclear military assets, because of their danger to civilian populations might undercut the effectiveness of a counterforce strategy both as a war-fighting strategy and as a deterrent.

There are doubts about the amount of collateral damage that would result from even a limited nuclear exchange. The main fear is that radioactive fallout could destroy or maim human beings thousands of miles away from the area of impact. Estimates about the number of people who would be affected and the extent to which they would be vary widely. Part of the problem with the conflicting estimates is that experts make widely different assumptions about targets and the number of nuclear explosions exchanged. Another part of the problem about estimating casualties from a nuclear exchange is that we are dealing with a fortunately unknown event. In any case, we may rationally assume that the radioactive fallout from even a limited nuclear exchange aimed at military targets would be likely to kill or maim hundreds of thousands, perhaps millions, of people worldwide.

There are other serious concerns about the ability to hold counterforce strategy to the limits its advocates envision. First, there is concern about $C.^3$ In a nuclear environment, local or regional commanders might either lose contact with superiors or decide to order nuclear strikes without authorization. Second, we cannot know whether or not Soviet leaders or Soviet commanders would observe the limits envisioned by proponents of a counterforce strategy. Third, we cannot be sure whether or not an American President would observe the limits envisioned by counterforce strategists in a crisis situation.

An additional and real concern is the fear that the Soviet leadership might interpret deployment of nuclear weapons to fit a counterforce strategy as a first-strike threat by the United States, and this might impel the Soviet leadership to use its own nuclear capacity in a first strike. Conversely, development of a credible American counterforce strategy might tempt an American President to strike first to "win" a nuclear war.

One can hardly be morally enthusiastic about the prospects of carrying out a counterforce strategy were deterrence to fail; any counterforce strategy threatens to kill or maim vast numbers of people. On the other hand, a strictly limited counterforce strategy may represent the minimum nuclear threat necessary to deter a Soviet nuclear attack, and unless such a Soviet attack can be deterred, the West would be hostage to Soviet blackmail. The ultimate prospect would be the possibility of total subjugation to the Soviet regime and the extinction of human freedom in the West. The proportionality we have to weigh is between the quantitative effects of a counterforce strategy on human life and the physical environment of human life were deterrence to fail, on the one hand, and the qualitative effects on human life were there no potentially credible nuclear deterrent, on the other.

The arguments that a counterforce strategy would not conform to the limits envisioned by proponents cannot be lightly dismissed. On the other hand, the arguments are admittedly conjectural. While C^3 is always a problem in war, and undoubtedly would be a greater problem in the context of nuclear war, the problem can be minimized. While we cannot be sure that the Soviet leadership would observe limits in a nuclear exchange, by the same token, we have no compelling reason to think that they would not. Indeed, it seems more likely that a Soviet leadership unwilling or indisposed to observe limits after a nuclear exchange took place would decide to launch an all-out nuclear strike to begin with. Antecedent clarity on U.S. intentions with respect to targets and limits might help marginally. Similarly, while an American President might go beyond the limits of a counterforce strategy in a crisis situation, there is no evident reason to suppose that he would not adhere to the previously debated and approved strategy.

The argument that a counterforce strategy would destabilize the current nuclear balance and invite a Soviet preemptive strike poses a real problem, but not necessarily an insurmountable one, if U.S. counterforce strategy remains consciously limited to a deterrent posture. The purpose of a counterforce—or any other—U.S. nuclear strategy is to deter a Soviet nuclear attack, not to defeat the Soviet Union in the event that deterrence were to fail. From that perspective, there is no point in targeting one more Soviet military asset than is necessary to deter the Soviets from initiating a nuclear war. If the United States were to restrict counterforce targeting to an absolute minimum, and if the United States could effectively communicate that message to the Soviet leadership by taking concrete steps unilaterally to reduce its surplus strategic nuclear arsenal, the Soviet leadership would have no motive to strike pre-emptively with nuclear weapons—unless, the Soviets themselves expected thereby to "win" a nuclear war.

The converse argument that a counterforce strategy might tempt an American President to strike first with nuclear weapons is subject to a similar analysis. If counterforce strategy restricts targeting to the minimum necessary to deter, an American President would hardly be tempted to initiate a nuclear war. Moreover, every American President in the nuclear age has indicated clearly an understanding of the horrors of nuclear war and a willingness to do everything possible to avoid it except in self-defense. There is no reason to suppose that a future American President would have any other understanding or willingness.

The central question about a counterforce strategy is whether or not it would be big enough to deter a Soviet nuclear attack and at the same time small enough to be morally proportionate to the death and destruction that would result if deterrence were to fail. Only a very limited U.S. nuclear response to a Soviet nuclear attack might narrowly satisfy the principle of proportionality. That limited targeting of Soviet military assets, however, might not be sufficient to deter the Soviet leadership. We simply cannot be sure what level of casualties the Soviet leadership would be prepared to accept. Of course, the Soviet leadership might be deterred by uncertainty whether or not the United States would adhere to its stated moral limits in a nuclear response to a Soviet nuclear attack.

Whether or not a limited counterforce strategy could be at the same time credible as a deterrent and morally proportional as a fighting strategy, in the event that deterrence were to fail, is a close question. But since both a countercity strategy and a countercontrol strategy would fail to satisfy the principle of proportionality, a counterforce strategy may be the only strategy with a chance of being at the same time credible as a deterrent and morally accepta-ble as a fighting strategy. This is where public debate and states-manship in the fullest moral sense are required.

Several points should be stressed about the morality of carrying out a counterforce strategy in the event that deterrence were to fail. First, we repeat that only limited nuclear targeting of Soviet mil-itary assets might satisfy the principle of proportionality. Second, carrying out a limited counterforce strategy would be a last-ditch attempt to avoid capitulation to Soviet nuclear blackmail with only a modest hope of success. Third—and here is the really bad news—once a limited nuclear response had been made, there would be no moral alternative to capitulation. This is indeed a hard statement to make, but moral integrity sometimes comes at a price. The price in this case could be the human freedom of the West. Socrates (Plato *Crito* 49 C, D) put the hierarchy of values quite simply and accurately: it is better to suffer injustice than to commit it.

The three nuclear strategies discussed so far—countercity, countercontrol, and counterforce—are designed to deter a Soviet nuclear first strike against the United States or a NATO ally. These strategies may also help to deter a Warsaw Pact conventional attack on Western Europe because the Soviet Union would be aware of the danger that the attack could escalate into nuclear war, but the strategies are not designed to deter such an attack. For that, NATO relies, in large part, on theatre and tactical nuclear weapons. To these, we now turn our attention.

Theatre and Tactical Weapons

To determine whether or not theatre or tactical nuclear targeting would satisfy the principle of proportionality is difficult. There are almost limitless possible mixes of nuclear and conventional weaponry in theatre defense, and NATO maintains an attitude of "creative ambiguity" with regard to its intentions on the use of theatre and tactical nuclear weapons.[10] The theory of "creative ambiguity" is that uncertainty adds to deterrence, and that uncertainty requires Warsaw Pact nations to dispose their troops and tanks in smaller concentrations to minimize the potential costs of a theatre or tactical nuclear response by NATO to a conventional attack. (Given the alternative of deterring a Soviet conventional attack on Western Europe by building up conventional defense forces, the prudence of relying on "creative ambiguity" in so potentially catastrophic a matter as the use of nuclear weapons is highly questionable. Clarity may contribute more to stability than ambiguity.)

We can, however, make a few general observations about the probable results and risks of NATO use of theatre or tactical nuclear weapons to defend Western Europe against a conventional attack by Warsaw Pact nations. Experts estimate that even a conventional defense of Western Europe would cause more death and destruction in populated areas than all of World War II.[11] To add destruction from theatre or tactical nuclear weapons (including radioactive fallout) to that from conventional weapons would, for all practical purposes, mean the end of Western European societies.

Initiation of theatre or tactical nuclear warfare would also risk an escalatory spiral from theatre or tactical to strategic nuclear warfare. In particular, there is uncertainty whether or not the Soviet leadership would be willing to observe thresholds and limits in nuclear warfare, especially if NATO appeared to be successfully defending Western Europe with theatre or tactical nuclear weapons. There is also doubt about the ability of NATO and Warsaw Pact commands to maintain control over a war possibly involving thousands of nuclear strikes. And, in the case of theatre nuclear weapons, the Soviet leadership might identify a theatre attack as the beginning of a strategic attack and respond accordingly against the United States. And, indeed, with more fuel and less payload, theatre nuclear weapons could in fact become strategic in their targeting.

Applying the principle of proportionality, we can draw certain tentative conclusions. The use of theatre or tactical nuclear weapons would cause disproportionate destruction because the Western societies defended would be practically destroyed. Conventional defense, of course, would also be highly destructive, but the use of theatre or tactical nuclear weapons in self-defense would transform a limited disaster into an unlimited catastrophe for Western Europe. In addition, the use of theatre or tactical nuclear weapons would disproportionately risk an escalation that might bring devastation to the United States, the Soviet Union, and neutral nations. These moral conclusions about proportionality would seem to apply as much to second as to first use of theatre or tactical nuclear weapons.[12]

Missile Defense

The last strategy to be considered is that of a defense against a Soviet ballistic missile attack.[13] Directed energy weapons, such as high-energy lasers and particle-beam weapons, could destroy attacking Soviet missiles in flight. The technologies for these weapons are still a decade away from development, but both the United States and the Soviet Union are devoting resources to research on the technologies.

Whatever the technical feasibility of a nuclear defense system, critics argue that any attempt by the United States to create such a system would radically destabilize the existing nuclear balance and thereby sharply increase the risk of nuclear war. With a near-perfect and fully operational nuclear defense system, the United States would be capable of attacking the Soviet Union without fear of reprisal. To prevent that capability, Soviet leaders might decide to strike before the system was operational. Conversely, according to the critics, possession of a nuclear defense system might tempt an overconfident American President to order a nuclear strike in a crisis situation.

Fear that an American President would resort to a nuclear strike against the Soviet Union on the basis of his confidence in a nuclear defense system seems irrational; any defect in the system would bring disaster on the United States as well as the Soviet Union in the event of nuclear war. Fear that the Soviet leadership might regard attempts by the United States to establish a near-perfect nuclear defense system as preparation for a nuclear attack, however, is more plausible. To put that fear to rest, it would be necessary for the two superpowers to cooperate in phasing nuclear defense systems in and offensive nuclear weapons out.

An ABM defense strategy is morally attractive because its purpose is to protect human life, not to take human life even indirectly. But the odds are against the technical feasibility of an absolutely "leak-proof" antimissile system. It would cost hundreds of billions of dollars that could be put to productive uses for human development. It would not protect against cruise or submarine-launched low-projectory missiles. And it is not at all likely that the U.S. and Soviet leaderships would agree to cooperate in the construction of ABM systems, that either superpower could afford to trust the other, or that information could be exchanged without endangering each superpower's defense. For these reasons, perfect nuclear defense is likely to remain an unrealizable aspiration.

Nuclear Deterrence and Nuclear Strategies

Having surveyed from a moral perspective the principal nuclear strategies available to the United States to deter a Soviet nuclear attack and NATO deployment of theatre and tactical nuclear weapons to deter a Soviet conventional attack on Western Europe, we can now return to the question of linkage between deterrence and willingness to use nuclear weaponry were deterrence to fail. If one were to conclude, as many do, that to carry out any nuclear war-fighting strategy would violate at least the principle of proportionality, and so that any use of nuclear weapons would be immoral, is it possible to deter Soviet nuclear blackmail while adhering strictly to moral principle?

The American Catholic bishops do not see any real possibility that the use of nuclear weapons could be kept within limits morally proportional to the death and destruction the weapons would cause.[14] At the same time, the bishops morally approve possession and (presumably) deployment of nuclear weapons to deter a Soviet nuclear attack, although the bishops hope such possession for deterrence purposes would be preparatory to nuclear disarmament.[15] The bishops do not articulate why nuclear weapons not intended for use in the event deterrence were to fail would deter the Soviets at all. Others, however, have articulated a theory to explain the paradox, and the theory is one of bluffing the Soviet leadership into good behavior.

Paul Ramsey explained the bluff theory nearly twenty years ago.[16] The United States would possess and deploy nuclear weapons as if the United States intended to retaliate massively to a nuclear attack.[17] Since, however, such retaliation would be immoral, the just American statesman would have no actual intention to do so. Ramsey argued that the Soviet leadership would not risk calling the U.S. bluff; the Soviets could not be sure whether or not American leaders would act morally in a crisis situation. The United States could thus preserve both moral integrity and national security.

Ramsey himself abandoned that position,[18] and defense analysts have generally ridiculed the theory. But the theory may not be so ridiculous. The Soviet leadership has demonstrated no disposition to credit U.S. good intentions. Second, the bellicose statements of U.S. Presidents and politicians may give them no reason to revise their opinions. Third, there is real reason to doubt what an American President would do in the event of nuclear attack or nuclear blackmail; there is no guarantee that any human being will act morally in a crisis. Fourth, the United States has at best a spotty record of acting morally when force is deemed necessary to protect national interests. Fifth, the Soviet leadership has almost always acted cautiously with respect to threatening U.S. or Western European territory. (The Cuban missile crisis is distinctive in that the U.S.S.R. then, arguably, undertook provocative action directly threatening U.S. territory.) And a limited counterforce strategy could give some beef to a purely bluff deterrent.

The bishops, citing experienced military and political leaders,[19] despair of the possibility of keeping nuclear warfare limited. They base that view on problems of C^3 uncertainty about Soviet observance of limits, and uncertainty about the ability of American leaders to stick to strict limits. As previously indicated, the chances that nuclear warfare would escape control or limits are substantial. Just-war doctrine, however, does not require certainty in calculating the balance between probable costs and probable gain in waging wars of self-defense.

Undoubtedly, any use of nuclear weapons raises the potential costs of waging war and includes risks that such a war would not remain controlled or limited. But solid probability, not certainty, is the norm here as elsewhere in moral decision making. Both the United States and the Soviet Union are conscious of their respective C^3 problems and trying to overcome them. The Soviet leadership shares with the United States an interest in limiting nuclear war, and leaders of the two powers could communicate through the "hot line" or the ambassadors of neutral nations during and after a strictly limited nuclear exchange. The bishops are surely correct, in view of the horrendous risks in nuclear war, that "the burden of proof remains on those who assert that meaningful limitation [of nuclear war] is possible,"[20] but proponents of a counterforce strategy might be able to nuance their position enough to satisfy the principle of proportionality and to develop effective ways and means to keep potential nuclear war strictly limited.

Nuclear Arms Control

The moral imperative to control the nuclear arms race and to dismantle existing nuclear arsenals is obvious. The problem is how to do so without jeopardizing the security of the United States and its allies, on the one hand, and the security of the Soviet Union and its allies, on the other. The history of nuclear arms control is a record of piecemeal agreements on controlling development of new nuclear weapons rather than dismantling existing ones. Neither side trusts the other, and the Soviet Union adamantly opposes on-site inspection by allied or neutral observers. All nuclear arms control agreements to date have been made with an eye to maintaining the MAD system of deterrence. There is a moral paradox here. The best prospects for future nuclear arms control agreements rest on maintaining the MAD balance of terror. But a

MAD system of deterrence would certainly (countercity strategy) or almost certainly (countercontrol strategy) be immoral if implemented in the event that deterrence were to fail.

This paradox poses a dilemma for future nuclear arms control negotiations. On the one hand, the prospects for any success in negotiating partial nuclear arms control agreements are best if the MAD deterrents are maintained. On the other, any unilateral movement by the United States away from a MAD deterrent in an attempt to meet just-war criteria could make nuclear arms control agreements more difficult if the Soviet leadership perceived the movement as indicative of a U.S. intention to "win" a nuclear war, perhaps even of an intention to strike first to do so.

A limited counterforce strategy, however, need not necessarily destabilize mutual security. A U.S. counterforce strategy might actually increase the security of the Soviet Union if the strategy followed strict limits on both the number of targets and the human environment of the targets. And a strictly limited U.S. counterforce strategy might increase the security of the United States and the West because the deterring threat would be more credible than MAD. From that perspective, both the U.S. and Soviet leaderships might have as much or more incentive to negotiate nuclear arms control agreements if the United States adopted a counterforce strategy than if it continued to follow a MAD strategy.

The United States, of course, would have to nuance carefully its counterforce strategy so that it would, in fact, be strictly limited, and the United States would have to communicate its intentions effectively to the Soviet leadership. One way to communicate effectively the strictly limited character of a counterforce strategy would be for the United States to dismantle unilaterally those parts of its nuclear arsenal which would become surplus under such a strategy.

The only way around the current deadlock on nuclear arms control is for both sides to accept risks to security in order to reduce the risk of nuclear war.[21] What is needed is less nuclear expertise and more political will. The present MAD systems give the illusion of security, and that is why both U.S. and Soviet leaderships are so reluctant to make any concessions that might weaken the systems. But the West and the Soviet Union would be as secure without nuclear weapons as they are now with nuclear weapons that neither wants to use. In fact, the two superpowers would be more secure without nuclear weapons because elimination of nuclear weapons would eliminate the possibility of nuclear war.

Nuclear arms control negotiations about existing weapons deadlock because neither side really wants to end the deadlock. We can
do nothing directly to alter the Soviet political will, but we can
invigorate our own.

Notes

1. See William V. O'Brien, "Just-War Doctrine in a Nuclear Context," *Theological Studies* 44 (September 1983):191–220. In the organization and structure of
this chapter, I follow his sophisticated discussion of the application of just-war
criteria to nuclear war.

2. This chapter focuses on the morality of Western reliance on nuclear weapons to deter Soviet attack and of possible Western use of nuclear weapons in the
event that deterrence were to fail. The general principles of discrimination and
proportionality developed here, however, should be applicable to other nations'
reliance on, and use of, such weapons.

3. Cf. Desmond Ball, "Can Nuclear War Be Controlled?," *Adelphi Paper*
no. 169 (London: Institute of Strategic Studies, 1981).

4. For an evaluation of the present state of U.S. C³ and the prospects for its
future improvement, see The Organization of the Joint Chiefs of Staff, *United
States Military Posture for Federal Year 1983* (Washington, D.C.: Government
Printing Office, 1982), pp. 81–83.

5. Leslie H. Gelb, "Is the Nuclear Threat Manageable?," *The New York
Times Magazine*, March 4, 1984, p. 35.

6. Ibid.

7. Letter of National Security Adviser, William P. Clark, to Joseph Cardinal
Bernardin, January 15, 1983. Clark there stated: "We do not threaten the existence
of Soviet civilization by threatening Soviet cities. Rather, we hold at risk the war-
making capability of the Soviet Union."

8. Some versions of countercontrol strategy call for the testing and deployment of space weapons to destroy Soviet satellites. Since satellites are the eyes and
ears of a control system, their elimination would weaken and disrupt Soviet
control over their nuclear weapons. At present, by all accounts, the Soviet Union
has only a small and unreliable antisatellite capacity. The U.S. is also developing
such a system. Although the most important Soviet and American satellites are
too far out in space to be destroyed by present weapons, that probably will no
longer be true in another decade.

Deployment of reliable antisatellite weapons in space would be very dangerous for two reasons. First, it would open up a new arms race, the dimensions of
which can only be imagined. Second, it would radically increase each side's fear
that the other was preparing for a first nuclear strike and might thus lead one or
the other to strike preemptively. Both the United States and the Soviet Union
have an interest in banning antisatellite electronic and laser weapons in space.
The sticking point, in the view of some, is verifiability. A first step toward a ban
on antisatellite weapons in space might be a ban on testing those weapons, which
should be as verifiable as other test bans. The present Soviet antisatellite capacity
is too unreliable and small to cause the U.S. much worry, and that capacity,
without testing, cannot be improved enough to become reliable.

9. Solly, Lord Zuckerman, *Nuclear Illusion and Reality* (London: Collins,
1982), p. 24.

10. See Laurence Martin, "Limited Nuclear War," in Michael Howard, ed., *Restraints on War: Studies in the Limitation of Armed Conflict* (Oxford: Oxford University Press, 1979), pp. 103–21.

11. O'Brien, p. 205.

12. The West need not be in a position where it must choose between the use of nuclear weapons and surrender in the face of a Warsaw Pact conventional attack. If the West were to increase its conventional defense capabilities, it would have no need to rely on a probably immoral and potentially suicidal nuclear war-fighting strategy to defend itself against a conventional attack. Improved conventional defense is not only a matter of quantitatively larger military forces and weaponry; improved conventional defense in line with just-war criteria is also a matter of qualitatively superior equipment and tactics to resist attack with minimum harm to civilian populations and with maximum control by commanding officers. Recent wars, e.g., between Great Britain and Argentina, and between Israel and Syria in Lebanon, demonstrate the effectiveness of sophisticated, non-nuclear weaponry in conventional wars. Tactical nuclear weapons may now offer little or no military benefits unobtainable by modern conventional weapons. The failure of the West, especially Western Europe, to redress the negative balance of conventional forces is only intelligible on an assumption of benign Soviet intentions. That may be wishful thinking.

13. Some proponents of an ABM defense system envision a three-tier "leak-proof" system: space-based lasers would eliminate 90% of attacking Soviet missiles as they rise from their launching pads; other weapons in space would eliminate 90% of those attacking missiles that survive the first-tier defense; and ground-based missiles would eliminate 90% of those attacking missiles that survive the second-tier defense. Thus, according to proponents, only one out of a thousand attacking Soviet missiles would survive the three tiers of defense.

Critics of the proposed system argue that it could be blinded by enemy destruction of vulnerable radars and satellites—the eyes and ears of the system. Second, critics argue that the system could be overwhelmed by the sheer number of incoming warheads, decoys, and chaff. Third, critics note that the system would not prevent cruise or submarine-launched low-projectory missiles from reaching homeland targets. Fourth, critics cite the astronomical financial cost of such a system.

14. National Conference of Catholic Bishops, *The Challenge of Peace: God's Promise and Our Response* (Washington, D.C.: U.S. Catholic Conference, 1983), ##157–61.

15. Ibid., ##186, 188. We focus in the text on the credibility of a nuclear deterrent unintended for use in the event deterrence were to fail. One may also ask whether or not the possession and deployment of nuclear weapons unintended for use would be moral. There is always a risk that any nuclear weapon might be accidentally detonated. Second, there is a risk that C^3 might break down. Third, there is a risk that the possession and deployment of nuclear weapons might tempt an American President to use them in a crisis situation. The American bishops thought assessment of these risks rested on "highly technical judgments about hypothetical events" and so declined to condemn possession and deployment of nuclear weapons exclusively for deterrence purposes (#192).

16. Paul Ramsey, *The Just War* (New York: Scribner's, 1968), pp. 249–58. J. Bryan Hehir, one of the most influential consultants of the American Catholic bishops in drafting the letter on nuclear war, seems to favor such an approach. See Robert A. Gessert and J. Bryan Hehir, *The New Nuclear Debate* (New York: Council on Religion and International Affairs, 1976), pp. 47–53, 66–69.

17. The American Catholic bishops are less precise and more circumspect; they say only that the United States should possess a "sufficiency" of nuclear weapons for deterrence purposes (#188).

18. "I now think that an input of deliberate ambiguity about the counter-people use of nuclear weapons is not possible unless it is (immorally) meant and not a very good idea in the first place." Paul Ramsey, "A Political Ethics Context for Strategic Thinking," In Morton A. Kaplan, ed., *Strategic Thinking and Its Moral Implications* (Chicago: University of Chicago Center for Policy Study, 1973), p. 142.

19. McGeorge Bundy, George F. Kennan, Robert S. McNamara, and Gerald Smith, "Nuclear Weapons and the Atlantic Alliance," *Foreign Affairs* 60 (Spring 1982):757; Arthur S. Collins, Jr., "Theatre Nuclear Warfare: The Battlefield," in John F. Reichart and Steven R. Sturn, eds., *American Defense Policy*, 5th ed. (Baltimore: Johns Hopkins, 1982), pp. 359-360; Harold Brown, *Department of Defense Annual Report, Federal Year 1979* (Washington, D.C.: Government Printing Office, 1978). Henry A. Kissinger, "Nuclear Weapons and the Peace Movement," *Washington Quarterly*, 5 (1982):31-39, is more optimistic.

20. *Challenge of Peace*, #159.

21. Nuclear arms control negotiations are complex and difficult. The United States and the Soviet Union have different mixes of nuclear weapons, and so limits on nuclear weaponry in which the United States has marked superiority need to be matched by limits on nuclear weaponry in which the Soviet Union has marked superiority if the two sides are to agree on specific nuclear arms control. Each side estimates the utility of nuclear weapons differently, and neither side trusts the other.

8

The Justice of Military Support or Intervention in Revolutionary Wars

In the last two chapters, we considered the application of just-war criteria to conventional and nuclear war. We now turn our attention to moral issues surrounding U.S. military support or intervention to put down reputed Marxist-Leninist revolutions in the Third World or to overthrow reputed Marxist-Leninist regimes already established there. As in the case of conventional war, we can conveniently divide the moral issues to be considered into two parts: the justice of the war decision to support or intervene and the justice of revolutionary-counterrevolutionary war conduct. Application of just-war criteria to revolutionary war is particularly difficult, not because the criteria are any less applicable to such war, but because factual elements on which moral judgments about revolutionary war depend are typically complex, hard to ascertain, and fluid. Nonetheless, we can line up the moral questions to ask.

Our focus will be theoretical. Inevitably, this writer and the reader will think in terms of Central America, the current Third-World revolutionary hotspot. The present administration is committed to military support for the government of El Salvador and has covertly and not-so-covertly supported rebels against the Sandinista government of Nicaragua. It is not the purpose of this chapter to pass moral judgment on those actions but to analyze the sort of factual determinations and moral assessments which would be necessary and sufficient to justify the actions. The general principles we develop here should be applicable to revolutions other than those in Central America and to revolutionaries other than Marxist-Leninists.

The Just War Decision to Support or Intervene

In our treatment of conventional and nuclear war, we had little difficulty identifying self-defense of a free society as a cause justifying war, provided that other just-war criteria had been satisfied. The same cannot be said about military support or intervention in behalf of one faction in a civil war. In fact, intervention in the domestic affairs of another nation is prima facie both immoral and contrary to the law of nations. It is prima facie immoral because every people has the right to determine for itself how its society should be organized. It is prima facie contrary to the law of nations because all nations have a common interest in peaceful conexistence.

Covert intervention, however, is a very old story in international affairs. The government of one nation naturally prefers to have friendly neighboring governments, and nations have frequently taken covert, including covert military, steps to assure that neighboring governments are friendly. Modern ideologies have added to the Machiavellian reasons for intervention in the affairs of other nations. When leaders of one or more nations see themselves as bearers of an ideological blueprint for reconstructing society worldwide, they will intervene whenever possible to make the blueprint a reality. And the more powerful targets of ideologically motivated intervention will then feel justified in intervening themselves, both to counter revolutionary forces threatening friendly regimes and to encourage revolutionary forces threatening hostile regimes. The French Revolution introduced ideology into relations among nations, and twentieth-century ideologies accentuated the trend. The ability of Nazi Germany and Fascist Italy to export ideological revolution was limited by the narrow racist and national bases of their ideologies. But Marxism-Leninism has no such narrow base, and so the Soviet Union and other Marxist-Leninist regimes have a far greater capacity to export their revolution worldwide.

The Just Cause of U.S. Security

In the context of the Marxist-Leninist ideology of world revolution, three basic arguments—arguments not always sufficiently distinguished in public policy debate—can be advanced to justify U.S. military support or intervention to quash Marxist-Leninist revolutions in Third-World nations or to topple Marxist-Leninist regimes in power there. The first argument appeals to the consequences of Marxist-Leninist regimes for U.S. or allied national security interests. Let us analyze the "domino" argument step-by-step.

A Communist government in control of a Caribbean island or a Central American nation and relying only on its own military and economic resources would hardly pose any direct threat to U.S. national security. But introduction of nuclear weapons or aircraft with missile capability there under Soviet control might raise legitimate U.S. security concerns. The Kennedy administration, for example, viewed the introduction of nuclear weapons under Soviet control into Cuba in 1962 as a serious threat to American security. Whether or not the presence of nuclear weapons in Cuba under Soviet control would have radically altered the existing MAD deterrance system—and the Kennedy view has been challenged—the point to note here is that the purported U.S. national security interest extended only to removal of the weapons from Cuba, not to overthrow of the Castro regime. The Kennedy administration acted on that assumption when it quarantined Cuba until the weapons were removed.

Aside from the potential threat posed by the presence of nuclear weapons or aircraft with missile capability on territory in close proximity to the United States, then, a Marxist-Leninist regime in the Caribbean or Central America would pose no direct threat to U.S. security. This is a fortiori true about Marxist-Leninist regimes further removed from U.S. territory. But it might be argued, and has been argued, that Marxist-Leninist regimes, especially in Central America, pose an indirect, long-term threat to U.S. security. If, for example, El Salvador were to fall under the domination of Marxist-Leninist revolutionaries, then Guatemala to the north and Honduras to the south would be threatened with similar revolutions. If Guatemala were to fall to Marxist-Leninist revolutionaries, Mexico would be threatened next. And if Mexico were to fall to Marxist-Leninist revolutionaries, potential or actual Soviet allies would be at the southern border of the United States. If Honduras were to fall under Marxist-Leninist domination, Costa Rica and then Panama would be threatened. And if Panama were to fall to revolutionaries sympathetic to the Soviet Union, the latter would control passage through the Panama Canal.

The "domino" theory involves a number of factual assumptions, and the validity of the theory as a whole and thus the reality of the threat to U.S. security depend on the plausibility of each assumption:

1. That Marxist-Leninists actually dominate a revolutionary movement or are likely to do so in the future.[1]

2. That successful Marxist-Leninist revolutionaries would attempt to export revolution to a neighboring country in the near or foreseeable future.[2]

3. That Marxist-Leninists successsful in one country would be successful in other countries.[3]

4. That the ultimate situation created by the "domino" effect would be highly threatening to U.S. or allied security interests.[4]

American statesmen considering intervention in such revolutionary situations must carefully assess facts and weigh arguments for and arguments against each of these assumptions.

The Just Cause of Human Rights

The second line of argument to justify U.S. military support of non-Communist regimes in the Third World is prima facie humanitarian; proponents argue that defense of human rights justifies military support of non-Communist regimes against Marxist-Leninist totalitarianism. The case of genocide is cited by way of analogy. Surely one nation is justified in intervening militarily in the affairs of another to stop genocide if the intervening nation has the power to do so. In like manner, so the argument runs, a powerful nation like the United States would be justified in supporting or intervening militarily to suppress Marxist-Leninist revolutions to prevent the irrevocable loss of an indigenous people's basic human freedoms and a possible bloodbath for opponents of the revolutionaries.

The humanitarian justification of military support or intervention in Third-World revolutions rests on assumptions that incumbent regimes are legitimate and just, and that revolutionary regimes would be illegitimate and unjust. Both legitimacy and justice are relative terms, not relative in the sense that they are purely subjective matters of personal opinion, but relative in the sense that legitimacy and justice represent ideals only more or less realized in the concrete order. What must be examined in particular cases is whether or not incumbent regimes are more legitimate and just than prospective revolutionary regimes would be.

Legitimacy and justice are distinct but closely connected concepts. While legitimacy refers only to popular support for a regime, and justice only to the fairness of a regime's political, economic, and social institutions and the resulting distribution of benefits and burdens among citizens, one concept typically involves the other. It is, of course, theoretically possible for a regime to enjoy broad popular support but to fail to conform substantially to standards of justice. The Nazi German regime would be one example of that. It is also possible for a regime to conform substantially to standards of justice but not to enjoy broad popular support. The pre-Soviet Kerensky regime in Russia might be an example of that. But typically legitimacy and justice go hand in hand. If a regime conforms substantially to standards of justice, it is likely to enjoy broad popular support, and if a regime is sensitive to popular demands, it is likely to conform substantially to standards of justice. Conversely, if a regime fails to conform substantially to standards of justice, it is not likely to enjoy broad popular support, and if a regime is not sensitive to popular demands, it is not likely to conform substantially to standards of justice.

There are three different levels of popular support. The first and most fundamental level of popular support is the recognition by individuals and groups that they constitute one people, one body politic. This recognition is frequently not the case in the Third World. In Black Africa, colonial powers drew up the present territorial lines on the basis of conquest and administrative convenience, and independent regimes there now have the difficult task of forging nations out of disparate tribes living within those territorial lines. In Central and South America, there are pockets of Indian populations, like the Mesquitos in Nicaragua, outside any mainstream national consciousness.

The second level of popular support is for the regime, that is, for the institutions and forms of government. In much or most of the Third World, the bulk of citizens have no deep loyalty to regimes because citizens do not participate politically in any real sense or feel that they have any economic stake in the regimes. Even where political processes are nominally democratic, voters may be intimidated, controlled, or simply bought.

The third level of popular support is for the currently governing personnel and their policies. As with support for the regime itself, many or most citizens in Third-World countries are at best passive with respect to support for incumbent government personnel and their policies. At the same time, Marxist-Leninist revolutionaries in the Third World may not have much popular support for the regime they propose to install or for the policies they would follow, at least at the outset of the revolution. The masses of citizens, by sheer weight of custom, are likely to be indifferent or fatalistic about their lot. The foremost task of counterrevolutionary and revolutionary alike is to move the minds and hearts of the indifferent masses to commitment.

Incumbent regimes in the Third World often fail to meet minimum standards of institutional and distributional jutsice. Political institutions may be authoritarian, with no tolerance of democratic dissent and little tolerance of religious freedom. The machinery of government may operate capriciously or ineffectively with respect to protection of human rights. Educational opportunities may be limited to the middle and upper classes. Economic institutions may keep the masses of agricultural workers in a state of serfdom while landlords reap large profits. The net result may be that the masses of citizens are condemned to a state of ignorance, poverty, and powerlessness.

Sophisticated defenders of U.S. military support for such regimes will admit that many non-Communist regimes fail, to a greater or lesser extent, to meet minimum standards of human rights and social justice. But they argue that the situation is not irreformable. By supporting presently oppressive regimes with economic and military aid, the United States will gain leverage to prod the regimes to institute political, economic, and social reforms. They admit that Marxist-Leninist regimes might improve the economic and social lot of the masses more rapidly, but they say that Marxist-Leninist regimes would be as politically oppressive as incumbent regimes. Moreover, with Soviet economic support and Marxist-Leninist indoctrination, the most basic human freedoms would be irretrievably lost. In short, military support for oppressive non-Communist regimes represents the lesser of two evils when the only alternative is a Marxist-Leninist regime.

Such a line of analysis rests on a number of factual assumptions and projections of future behavior. It assumes that revolutionaries are hierarchically organized, and that Marxist-Leninists control the group. Second, it assumes that Marxist-Leninists will retain complete control of the group after a successful revolution. Third, it assumes that successful revolutionaries will apply Marxist-Leninist ideology according to the Soviet or Maoist model. And with respect to the possibility of reforming the incumbent regime, it assumes that U.S. support will induce reform.

These assumptions and projections will have varying degrees of probability in concrete cases, and pessimism about the character of both revolutionaries and counterrevolutionaries may well be prudent. If so, then the choice will be between military support for an oppressive regime and a hands-off policy that will most likely result in a Marxist-Leninist victory.

Such a choice would be brutally stark. Some would argue that the long-term possibility of a free and just society is preferable to short-term advances in economic and social justice at the cost of quasi-permanent loss of freedom.[5] Others would argue to the contrary that rectifying present economic and social injustices is preferable to only an abstract future possibility of justice with freedom. Neither choice is morally appealing.

One other possibility should be noted. If the United States were to refuse military support to an oppressive regime threatened by Marxist-Leninist revolution, unless the regime radically reformed, the regime might have greater incentive to reform than if the United States gave military support without conditions. And if a regime were to refuse to reform sufficiently to obtain U.S. military support and were later to fall, the lesson might be salutary for leaders of other oppressive regimes confronting Marxist-Leninist revolutionaries now or in the future.

The Just Cause of Resisting Foreign Intervention

The third line of argument to justify U.S. military support for intervention in Third-World revolutionary wars reasons that such support or intervention is necessary to counter outside support of Marxist-Leninist revolutionaries by the Soviet Union, China, Cuba, Vietnam, or other Communist regimes. This line of argument combines elements from the U.S. interest in national security and the humanitarian interest in indigenous peoples determining for themselves how their societies will be organized. Its strength as a moral argument depends on those elements, and so our analysis will not differ greatly from analyses of the first two lines of argument to justify U.S. military support or intervention in Third-World revolutionary wars.

Outside support of Marxist-Leninist revolutionaries would violate the right of a people to determine for itself how to organize a free and just society. But to focus on outside support or intervention in behalf of Marxist-Leninist revolutionaries distracts attention from the central problem. No popularly based revolution has a chance of success, with or without foreign support, unless large segments of the population are indifferent or hostile to the incumbent regime. Outside support or intervention would only become a critical factor for revolutionary success if the support were to be the vehicle whereby a small minority could overwhelm a majority loyal to the incumbent regime. Whether or not outside support is a critical factor will depend on the quantity and quality of the outside support.

We can distinguish two polar situations in which outside regimes support local Marxist-Leninist revolutionaries. In the first situation, outside regimes supply only arms, ammunition, and military advisers to local revolutionaries. If that were the case, outside support might keep the revolution going, but, by itself, it could not achieve victory. The success of the revolution would ultimately depend on two local factors: sufficient military personnel and sufficient popular support or indifference. Outside support would undoubtedly help local revolutionaries, but local military activity and local popular support would remain the critical factors. (One might also note parenthetically that local revolutionaries often rely as much on arms and ammunition captured from government forces as on those supplied by foreign sources.)

In the second situation, outside regimes supply military forces to fight alongside Marxist-Leninist revolutionaries. If that were the case, outside support from Marxist-Leninist regimes would indeed be a critical factor in a revolutionary war, and U.S. counter-support for the incumbent regime might be justified on the humanitarian ground of helping a people to defend itself from outside intervention—provided that the incumbent regime be relatively legitimate and just.

Other Just War-Decision Criteria

The right intention in military support or intervention in behalf of incumbent regimes, as in all war, is to right wrong, not to satisfy national pride. In the absence of a legitimate U.S. national security interest, the right or wrong of revolutionary wars concerns the welfare of the indigenous population. The right of waging revolutionary or counterrevolutionary war involves the freedom of a people and the justice of its regime; the wrong of waging revolutionary or counterrevolutionary war is unfreedom and injustice. It is no part of right intention to vindicate American pride, to protect American economic interests, or to score a triumph over Marxist-Leninists. To maintain right intention while waging conventional war is difficult enough, but to maintain right intention while waging revolutionary or counterrevolutionary war will be still more difficult. Revolutionary war is typically "dirty," and brutality by one side tends to trigger brutality by the other.

Article I (sect. 8, cl. 1) of the U. S. Constitution requires that Congress authorize public expenditures and appropriate the funds. The presidential role under the Constitution, besides those of recommending and signing bills, is limited to administering the expenditures authorized and the funds appropriated. With respect to expenditures and funds to provide military aid to foreign governments, however, the political fact, as opposed to the legal theory, is that the President is the principal decision maker.

The President controls the information on the basis of which Congress acts, and only Congressmen on select committees have limited access to that information. Congressmen on those committees are likely to defer to the President's judgment on military support, and ordinary Congressmen to their colleagues on the special committees. After appropriation of funds, Congress exercises little oversight, and the President has a relatively free hand. If the President's control of information helps to make him the most influential maker of decisions whether or not to provide military support to help suppress Third-World revolutions, control of information by military and C.I.A. personnel on the spot helps to give them in turn a powerful voice in deciding how much support is needed, and how the funds are spent. There are thus real problems about supporting counterrevolutionary (or revolutionary) wars from the just-war perspective of control by civilian and constitutional authority.

The constitutional authority of the President to send U.S. armed forces into warring areas without Congressional authorization to do so is ambiguous. On the one hand, the President is Commander in Chief of the armed forces. On the other, Congress alone can declare war. Commentators generally agree that the President may constitutionally deploy American armed forces to repel an attack on American territory or an attack on American forces in international sea or air space without a declaration of war or other specific congressional authorization. Similarly, commentators generally agree that the President may deploy American forces abroad to protect American nationals without specific congressional authorization.

The War Powers Resolution of 1974 attempted to limit the President's power to make ad hoc decisions to deploy or engage American forces in hostile situations abroad. Congress feared that such presidential decisions could lead the nation into open-ended hostilities and so required him to consult with Congrss beforehand and to obtain approval from Congress subsequently whenever engagement of American forces goes beyond 60 to 90 days. Every President from Nixon to Reagan has maintained that the War Powers Resolution unconstitutionally interferes with the President's role as Commander in Chief. Since there has been no authoritative decision by the Supreme Court on the point, the question is still an open one. In any case, the War Powers Resolution has not substantially curtailed the President's freedom of action because he controls information about foreign events, and because Congress tends to defer to his judgments about them.

An assessment of the just-war criterion of competent authority must also examine the assumption that it is the primary right and duty of the United States to protect Third-World peoples from Marxist-Leninist revolutionary takeovers. There are alternate candidates for that role. The United Nations, for example, might conceivably intervene. Article 42 of the U.N. charter empowers the Security Council to use the armed forces of member states "to maintain or restore international peace and security." U. N. intervention is hardly likely in the case of Third-World Marxist-Leninist revolutions since the Soviet Union and China, as permanent members, can veto Security Council resolutions. Regional associations also might intervene, but this is unlikely in most revolutionary hotspots for a number of reasons. Outside of NATO, there are few, if any, effective regional defense associations. Nations of a region may be reluctant to intervene, and they may not have sufficient force to be effective in any case. What the U.N. Secretary General and regional diplomats can do is to offer their services to mediate a compromise mutually acceptable to the warring parties.

The problem of proportionality is closely related to the problem of just cause itself. The amount of death and destruction that would be morally tolerable in helping to wage a counterrevolutionary (or revolutionary) war depends, in part, on how just is the cause of war. Hope for a free and relatively just society in the future might marginally justify supporting the counterrevolutionary war of a presently oppressive regime, but such a cause would hardly justify very large-scale death and destruction. On the other hand, helping to defend a presently free and relatively just society would justify considerable death and destruction in the course of the war. Even the defense of a free and relatively just society, however, would no more justify support for a war devastating a people and its social and economic fabric than defense of Western Europe against Soviet attack would justify nuclear warfare which would produce such results. Exactly what threshold of death and destruction short of that extreme would constitute the morally tolerable limit in waging war in defense of free and relatively just regimes against Marxist-Leninist revolutionaries is a matter for practical judgment.

The level of death and destruction likely to result from revolutionary war is closely linked to the relative legitimacy of the incumbent regime. Where only a small percentage of the indigenous population is loyal to a regime, and most of the population at best indifferent, counterrevolutionary warfare is likely to be very destructive. But where a large percentage of the indigenous population is loyal to the regime, and only a small percentage hostile, counterrevolutionary war is not likely to be nearly so destructive—provided outside armed forces do not intervene in support of the revolutionaries, in which case a revolutionary war would be transformed into a conventional war.

Our focus in this chapter has been on revolutionary wars in the Third World. Another revolutionary situation to be considered is one in which a people of the Second World, that is, Eastern Europe, revolted against their Communist rulers and called upon the United States for military support. There would be little doubt about the justice of that cause. There would at the same time be little doubt that overt U.S. military support of the revolutionaries would trigger a Soviet response, possibly a nuclear response, involving massive death and destruction in the West as well as the East.

Another just-war criterion requires that just warriors have a reasonable hope of success. This will depend, in part, on the military situation and the stage of revolutionary activity. But success in suppressing Marxist-Leninist revolutionaries depends in larger part on the political situation, that is, on the quantitative and qualitative support that the native population accords the incumbent regime. Where there is little indigenous support for the regime, large-scale military support would not be likely to win a clear-cut victory for the regime, at least not without a very long war.[6] And popular support for a regime, in turn, is likely to depend on popular perception of the regime's justice.

The last criterion for a just war decision to be considered is that requiring just warriors to exhaust peaceful means to resolve the conflict before they resort to war and to continue to search for peaceful means after they resort to war. It is difficult to apply this criterion to revolutionary wars under Marxist-Leninist leadership, as the Marxist-Leninist ideology of class warfare does not allow for any permanent peace between Marxist-Leninist and non-Marxist-Leninist elements in society. Nevertheless, within that ideological framework, Marxist-Leninists might accept a temporary truce in the on-going class war for pragmatic reasons, and the truce might be acceptable to non-Marxist-Leninists as well.

If a regime enjoys broad popular support and is broadly perceived as just, revolutionaries might find it in their interest to lay down arms temporarily in return for amnesty and admittance to the political process. If, however, a regime enjoys little popular support and is broadly perceived as unjust, revolutionaries would be unlikely to find it in their interest to lay down arms when the conditions for successful revolution seem so ripe. Thus, political, economic, and social reforms by incumbent regimes are central to the possibility of negotiating peace with Marxist-Leninist revolutionaries without capitulation.

In in-between situations, a regime will be struggling to achieve broader popular support and broader popular perception of the regime's justice. In that case, counterrevolutionary war would be necessary to buy time for reforms to increase popular support for the regime, and Marxist-Leninist revolutionaries would be likely to find it to their interest to step up military activities and to sabotage the reforms. But Marxist-Leninist revolutionaries might be persuaded to accept some form of "power sharing." Short of de facto partition, however, the mutual distrust between Marxist-Leninists and non-Marxist-Leninists makes "power sharing" unlikely and risky for the non-Marxist-Leninists in any case.

Regional groups might be able to mediate agreement on limiting outside support to both sides, exchanges of prisoners, and rules of war conduct. Bilateral agreements between the United States and Marxist-Leninist regimes to limit outside military support might also be possible. Or the United States might agree to forego military support for revolutionaries against Marxist-Leninist regimes in countries like Afghanistan in return for agreement by the Soviet Union to forego military support of revolutionaries against non-Marxist-Leninist regimes in countries like El Salvador.

All of these possibilities of a negotiated settlement must be explored if just warriors are to meet the requirement to exhaust peaceful means.

The Just Conduct
of Revolutionary-Counterrevolutionary War

Just-war theory requires that war be waged discriminately, that is, against the enemy-in-arms and his waging of war, not against noncombatants unconnected with the waging of war. The principle of discrimination is difficult to apply to any modern warfare. Since the French Revolution, Western nations waging conventional war have mobilized all or most of their resources behind the war effort. In the circumstances of modern war, it is difficult to distinguish between the guilty enemy and the innocent enemy. It is still more difficult in the circumstances of modern revolutionary war to distinguish between combatants connected with the waging of war, on the one hand, and those non-combatants unconnected with the waging of war, on the other.

Just war presupposes that the enemy's cause is unjust and justifies only the killing of those enemy who are committing injustice. The principle clearly applies to, and condemns, the acts of terrorism so characteristic of modern revolutionary wars, whether practiced by revolutionaries or counterrevolutionaries. Terrorism is a deliberate policy of inflicting death and destruction on civilian populations in order to undermine enemy morale and willingness to fight. Terrorism makes no attempt to distinguish the guilty enemy from the innocent enemy, and, in fact, terrorism will succeed in its objective only to the extent that innocent enemy are killed. Terrorism rests on the principle that the end justifies all killing in war; just-war theory rests on the principle that no end can justify directly killing those committing no wrong.

The United States may not be able to control entirely the behavior of a regime which the United States is militarily supporting, although the United States can and should condemn acts of terrorism committed by revolutionaries and counterrevolutionaries alike. But the United States is the master of its own behavior and that of those officially representing it in a foreign country. The United States can and should refuse to participate in acts of terrorism and punish any U.S. personnel who participate in such acts.

An isolated act of terrorism would not necessarily render unjust continued U.S. military support for a regime whose personnel were responsible for the act. Allied obliteration bombing of Dresden and other German cities toward the end of World War II violated the principle of discrimination, as did the U.S. nuclear

bombing of Hiroshima and Nagasaki. Since those acts were relatively untypical of overall Allied conduct of the war, most commentators were at the time, and continue to be, unwilling to
condemn the entire U.S. and Allied war conduct against Germany
and Japan as unjust. But the systematic practice of terrorism by a
U.S.-supported regime would render a counterrevolutionary war
unjust. Under such circumstances, continued U.S. military support of the regime could not be morally justified.

The temptation for both sides to practice terrorism is great.
Revolutionaries are tempted to do so in order to discourage local
populations from reporting the location and movements of the
revolutionary forces. Counterrevolutionary forces are tempted to
do so in order to punish local populations friendly to the rebels
and to induce future cooperation. Moreover, the frustration of
fighting a counterrevolutionary war may lead government troops
to commit atrocities like that at My Lai. None of these considerations can justify terrorist acts.

Waging modern counterrevolutionary (or revolutionary) war,
even without recourse to terrorism, raises problems of proportionality. Guerrilla warfare is a deadly game of hide-and-seek. The
exact location of revolutionaries is rarely known and is transitory
in any case. This means that aerial bombing and artillery fire
attempting to destroy revolutionary personnel and bases are likely
to cause widespread civilian casualties. From a moral perspective,
the incidental death of civilians can be justified only if proportional to the importance of the military target and the probability
that the attack will achieve its objective. If counterrevolutionary
forces know only the province or region where revolutionaries
train their forces, there would seem to be no morally acceptable
proportion between almost random aerial bombing or artillery fire
and the likelihood of achieving the military objective. And satura-
tion bombing or artillery fire might result in civilian casualties
morally disproportionate to the importance of a military target,
however high the probability that the military target would
thereby be destroyed.

Counterrevolutionary war tactics often involve aerial bombing
of agricultural crops thought to be destined for consumption by
revolutionary forces. This too raises an issue of proportionality.
Even if such bombings were to result in few civilian casualties, it
might destroy the means of civilian subsistence in the region.
Again, the importance of the objective and the likelihood of achieving it have to be weighed against the costs to the civilian popula-

tion. Both the importance of denying food supplies to revolutionary forces and the likelihood of doing so by the bombing can be exaggerated. Revolutionaries may alter their diet, their sources of food supplies, or the location of their forces. (The costs of bombing to the food supplies of civilians, of course, may also be exaggerated for similar reasons.)

Just-war theory requires that counterrevolutionary war be waged discriminately and proportionally. Pragmatic considerations counsel the same policy. It cannot be overemphasized that the objectives of armed conflicts between counterrevolutionaries and Marxist-Leninist revolutionaries are not simply military but also and fundamentally political. Neither counterrevolutionary nor revolutionary forces, with or without outside assistance, can win military victory without popular support or at least popular acquiescence. From that perspective, the principles of discrimination and proportionality in waging counterrevolutionary war are pragmatic counsels as well as moral injunctions. Indiscriminate and disproportional warfare would alienate the very public whose support or acquiescence is essential to military victory.

Notes

1. American statesmen must carefully distinguish revolutionary rhetoric from revolutionary reality and revolutionary parts from revolutionary wholes. Those who profess themselves Marxists may not be Leninists or not thoroughly such, and doctrinaire Marxist-Leninists may be too few or too incohesive to dominate a revolutionary movement.

2. Supporters of American intervention in Vietnam predicted large-scale "domino" effects in Southeast Asia if the Hanoi regime and its Vietcong allies triumphed. It is now ten years since the collapse of the Saigon regime, and the predicted effects have not generally taken place. The Communist regime in a unified Vietnam did occupy Cambodia. The regime which Communist Vietnam deposed there (the Khmer Rouge), however, was itself Communist, and Communist China was the perceived threat.

3. Neighboring non-Communist regimes may enjoy enough legitimacy to suppress Marxist-Leninist revolutions without American aid, provided that the revolutionaries receive no substantial external military support.

4. Even the presence of a conventional army on the banks of the Rio Grande, for example, would hardly pose a substantial threat to U.S. security. Moreover, Marxist-Leninist revolutionaries come to power might not be or remain subject to the beck and call of other Marxist-Leninist regimes.

5. In view of the record in Eastern Europe, of course, it may not be the case that presently oppressed citizens of the Third World would be economically and socially better off under a Marxist-Leninist regime.

6. Large-scale U.S. military assistance to a regime lacking broad popular support might, paradoxically, help the rebels in some ways. First, military supplies delivered to the regime would become a target of ambushes, and so a sizeable portion of the supplies might end up in revolutionary hands. Second, large-scale military support would increase the tempo of the war, and the resulting death and destruction might move masses of citizens from indifference to hostility to the regime. Third, the visible presence of U.S. advisers and military personnel might undercut whatever legitimacy the regime enjoys and foster anti-American sentiment. Lastly, the influx of money and supplies from the United States might cause runaway inflation or be siphoned off into the hands of corrupt officials and so engender resentment against the regime and the United States.

A Bibliographical Note

Chapter 1

The central text of Aristotle's ethics, of course, is the *Nicomachean Ethics*. Martin Ostwald provides a literal and literate translation, with introduction and notes (Indianapolis: Bobbs-Merrill, 1962). One of the best commentaries is that by Aquinas. See Thomas Aquinas, *Commentary on the Nicomachean Ethics*, trans. C. I. Litzinger, 2 vols. (Chicago: Regnery, 1964). René A. Gauthier, *La morale d'Aristote* (Paris: Presses Univeraires de France, 1958), offers an insightful and provocative interpretation. Also see William F. Hardie, *Aristotle's Ethical Theory* (Oxford: Clarendon Press, 1968). Frederick C. Copleston, S.J., *A History of Philosophy*, 9 vols. (Garden City, N.Y.: Image, 1962–77), 1, part 1: 74–91, provides an excellent introduction to Aristotle's ethical theory. Harry V. Jaffa, *Thomism and Aristotelianism* (Chicago: University of Chicago Press, 1952), carefully distinguishes Thomistic from Aristotelian ethical theory. Unfortunately, Jaffa tends to discount the rationality of Aquinas's distinctive theory on the ground that it is admittedly informed by faith. Aquinas's theory, however, claims to be accessible to reason and deserves to be evaluated as such.

The central text of Aquinas's moral theory is ST 1-2. 90–105. The standard translation is that of the English Dominican Fathers, *The Summa Theologica of St. Thomas Aquinas*, 2d ed. rev.; 3 vols. (New York: Benziger, 1947), 1:993–1103. That translation is also available in *Basic Writings of Thomas Aquinas*, ed. Anton C. Pegis, 2 vols. (New York: Random House, 1945), 2:742–948. For commentaries, see Jacques Maritain, *Man and the State* (Chicago: University of Chicago Press, 1951), pp. 84–97, Andrew C. Varga, S.J., *On Being Human* (New York: Paulist Press, 1978), especially pp. 61–69, 97–105, and Copleston, 2, part 2:118–31. Ross A. Armstrong, *Primary and Secondary Precepts in Thomistic Natural Law Teaching* (The Hague: Nijhoff, 1966), explicates thoroughly not only Aquinas's distinction between primary and secondary precepts but also his distinction between proximate and remote secondary precepts. Patrick Lee, "Permanence of the Ten Commandments: St. Thomas and His Modern Commentators," *Theological Studies* 42 (September 1981):422–43, conclusively demonstrates that Aquinas did not regard divine commands to kill nonaggressors as dispensations from natural law but rather regarded such divinely commanded acts as essentially different from those prohibited by natural law.

On contemporary Catholic consequentialism, the works of Richard A. McCormick have been cited in the text. Also see Charles E. Curran, ed., *Absolutes in Moral Theology* (Washington, D.C.: Corpus, 1968). On the "basic values" approach, see John Finnis, *Natural Law and Natural Rights* (Oxford: Clarendon Press, 1980), and Germain G. Grisez and Russell B. Shaw, *Beyond the New Morality: The Responsibilities of Freedom* (Notre Dame, Ind.: University of Notre Dame Press, 1974). For defenses of the traditional Thomistic position, see William E. May, *Becoming Human: An Introduction to Christian Ethics* (Dayton: Pflaum, 1975), and Ralph M. McInerny, *Ethica Thomistica: The Moral Philosophy of Thomas Aquinas* (Washington, D.C.: Catholic University Press, 1982).

On the New Testament notion of conscience, see Claude A. Pierce, *Conscience in the New Testament* (London: SCM Press, 1955).

Chapter 2

T. A. Sinclair offers a good translation, with introduction, of Aristotle's *Politics* (Harmondsworth, Eng.: Penguin, 1962). Sir Ernest Barker, trans., *The Politics of Aristotle*, with Introduction and Notes (New York: Oxford University Press, 1958), provides an excellent commentary. Aquinas's *Commentary on the Politics* is also excellent but unavailable in English translation. Also see Harry V. Jaffa, "Aristotle," *History of Political Philosophy*, eds. Leo Strauss and Joseph Cropsey, 2d ed. (Chicago: Rand, McNally, 1972), pp. 64–129.

Dino Bigongiari, ed., *The Political Ideas of Thomas Aquinas*, with Introduction (New York: Hafner, 1953), includes the English Dominican Fathers' translation of the most relevant sections of the *Summa theologica* and the Phelan-Esclimann translation of the most important sections of the *De regno*. For commentaries, see Thomas Gilbey, O.P., *The Political Thought of Thomas Aquinas* (Chicago: University of Chicago Press, 1958), and Alessandro P. d'Entrèves, *The Notion of the State: An Introduction to Legal Theory* (Oxford: Clarendon Press, 1967). For shorter treatments, see Copleston, 2, part 2:132–43, and Robert W. and Alexander J. Carlyle, *A History of Medieval Political Thought in the West*, 6 vols. (Edinburgh: Blackwood, 1903–36), 5:89–97.

The text of Thomas Hobbes's *Leviathan* is widely available. For longer commentaries, see Leo Strauss, *The Political Philosophy of Hobbes: Its Basis and Its Genesis*, trans. Elsa M. Sinclair (Oxford: Clarendon Press, 1956), J. Howard Warrender, *The Political Philosophy of Hobbes: His Theory of Obligation* (Oxford: Clarendon Press, 1957), and David Gauthier, *The Logic of Leviathan: The Moral and Political Theory of Thomas Hobbes* (New York: Oxford University Press, 1969). For a shorter treatment, see Copleston, 5, part 1:41–60.

Peter Laslett critically edited the text of John Locke's *Two Treatises of Government: A Critical Edition with an Introduction and Apparatus Criticus*, 2d ed. (London: Cambridge University Press, 1967). The text of *The Second Treatise* is widely available in paperback editions. See, for example, John Locke, *The Second Treatise of Government*, ed. Thomas P. Peardon, with Introduction (New York: Liberal Arts Press, 1952). Maurice Cranston, *John Locke: A Biography* (New York: Macmillan, 1957), provides an excellent biography. On the linkage of Locke's epistemology to his moral theory, see John Colman, *John Locke's Moral Philosophy* (Edinburgh: Edinburgh University Press, 1983). On the role of consent in Locke's political theory, see Jules Steinberg, *Locke, Rousseau, and the Idea of Consent* (Westport, Conn.: Greenwood, 1978). For a short, general treatment, see Copleston, 5, part 1:133–52. On Hobbes's and Locke's individualism, see C. B. Macpherson, *The Political Theory of Possessive Individualism: Hobbes to Locke* (Oxford: Clarendon Press, 1962).

John E. Elliott, ed., *Marx and Engels on Economics, Politics, and Society: Essential Readings with Editorial Commentary* (New York: Scott, Foreman, 1981), provides a large collection of the writings of Marx and Engels. Robert C. Tucker, ed., *Marx-Engels Reader*, 2d ed. (New York: Norton, 1978), provides a shorter collection. Tom Bottomore, ed., *Karl Marx* (Oxford: Blackwell, 1979), offers interpretative essays on Marx's theory. Also see David McLellan, *The Thought of Karl Marx: An Introduction* (London, Macmillan, 1971). Tom Bottomore, ed., *Modern Interpretations of Marx* (Oxford: Blackwell, 1981), offers reassessments of Marx's thought in recent decades. For explanations of Communist theory and practice in the Soviet Union, see Victor Ferkiss, *Communism Today: Belief and Practice* (New York: Paulist Press, 1962); Lester Kolakowski, *Main Currents of Marxism: Its Rise, Growth, and Dissolution*, trans. P. S. Falla, 3 vols. (New York: Oxford University Press, 1981); Gustav A. Wetter, S.J., *Dialectical Materialism: A Historical and Systematic Survey of Philosophy in the Soviet Union*, trans. Peter Heath (New York: Praeger, 1958). The works of Kolakowski and Wetter are comprehensive.

Chapter 3

The most important contemporary work on general principles of justice is that of John Rawls, *A Theory of Justice* (Cambridge, Mass.: Harvard University Press, 1971). I offer a comprehensive review of U.S. Supreme Court decisions involving conscientious objection up to 1972 in *Private Conscience and Public Law: The American Experience* (New York: Fordham University Press, 1972). For decisions after 1972, see my "Church and State in the United States," *New Catholic Encyclopedia*, 1979 Supplement, and my annual "Supreme Court Roundup," *Thought*, from 1979 to date. Also see Michael Walzer, *Obligations: Essays on Disobedience, War, and Citizenship* (Cambridge, Mass,: Harvard University Press, 1982).

Chapter 4

On the general topic of public philosophy and civic virtue, mention has already been made in the text of the works of Sir Ernest Barker, *Traditions of Civility* (Cambridge, Eng.: Cambridge University Press, 1948), Walter Lippman, *The Public Philosophy* (Boston: Little, Brown, 1955), and John C. Murray, S.J., *We Hold These Truths* (New York: Sheed and Ward, 1960). On legal aspects of the abortion question, see John H. Ely, "The Wages of Crying Wolf: A Comment on *Roe v. Wade*,"*Yale Law Journal* 82 (May 1973):920–49, and Louis Henkin, "Privacy and Autonomy," *Columbia Law Review* 74 (December 1974):1410–33. On philosophical aspects of abortion and other cases of life-taking, see Philip E. Devine, *The Ethics of Homicide* (Ithaca: Cornell University Press, 1978). On the historical development of the Catholic theological position on abortion, see John R. Connery, *Abortion: The Development of the Roman Catholic Perspective* (Chicago: Loyola University Press, 1977).

Chapter 5

The relevant tests of Aquinas are cited in the text, as is the authoritative article of Anthony Parel, "The Thomistic Theory of Property, Regime, and the Good Life," *Calgary Aquinas Studies*, ed. Anthony Parel (Toronto: Pontifical Institute of Medieval Studies, 1978), pp. 77–104. On the Thomistic virtue of justice and its divisions, see Joseph Pieper, *Justice*, trans. Lawrence E. Lynch (New York: Pantheon, 1955), William F. Drummond, S. J., *Social Justice* (Milwaukee: Bruce,

1955), and Arthur F. McKee, "What is 'Distributive Justice'?," *Review of Social Economy* 39 (April 1981):1–17. The teachings of the modern papacy are highly relevant: Leo XIII, "Rerum Novarum," *Acta Sanctae Sedis* 23 (1890–91):643–70; Pius XI, "Quadrigesimo Anno," *Acta Apostolicae Sedis* 23 (1931):177–228; John XXIII, "Mater et Magistra," *Acta Apostolicae Sedis* 53 (1961):401–64; Paul VI, "Populorum Progressio," *Acta Apostolicae Sedis* 59 (1967):257–99; Paul VI, "Octogesima Adveniens," *Acta Apostolicae Sedis* 63 (1971):401–41; John Paul II, "Laborem Exercens," *Acta Apostolicae Sedis* 73 (1981):577–647. (English translations of the latter are available from America and Paulist Presses.) *Mater et Magistra* and *Laborem Exercens* are particularly recommended. Also see the American Catholic bishops' *Pastoral Letter on Catholic Social Teaching and the U.S. Economy* (Washington, D.C.: U.S. Catholic Conference, 1985). For a comprehensive commentary on Catholic social thought up to 1965, see Rodger Charles, S.J., with Drostan Maclaren, O.P., *The Social Teaching of Vatican II* (San Francisco: Ignatius Press, 1982).

David Miller, *Social Justice* (Oxford: Clarendon Press, 1976), distinguishes distributive justice in traditional societies from that in modern societies. For an unequivocally laissez-faire view of distributive justice, see Robert Nozick, *Anarchy, State, and Utopia* (New York: Basic Books, 1974). On some shortcomings of capitalist practice from the perspective of capitalist theory, consult Lester Thurow, *Dangerous Currents* (New York: Random House, 1984). For structural critiques of capitalism, see the following: Richard C. Edwards, Michael Reich, and Thomas Weisskopf, *The Capitalist System*, 2d ed. (Englewood Cliffs, N.J.: Prentice-Hall, 1978); Kai Nielsen, *Equality and Liberty: A Defense of Radical Egalitarianism* (Totowa, N.J.: Rowman and Allanheld, 1984); Nicos Poulanzas, *Political Power and Social Classes* (London: New Left Books, 1973); Jurgen Habermas, *Knowledge and Human Interests*, trans. Jeremy J. Shapiro (Boston: Beacon Press, 1971); Jurgen Habermas, *Communication and the Evolution of Society*, trans. Thomas McCarthy (Boston: Beacon Press, 1979). For a sample welfare capitalist view, see Lester Thurow, *Zero-Sum Society* (New York: Basic Books, 1980). For a critique of the impact of the international capitalist system on the Third-World poor, see Francis Moore Lappé and Joseph Collins, *Food First: Beyond the Myth of Scarcity* (New York: Ballantine, 1979), and James B. McGinnis, *Bread and Justice: Toward a New International Economic Order* (New York: Paulist Press, 1979).

The principal works of liberation theologians Gustavo Gutiérrez, Juan L. Segundo, José M. Bonino, Hugh Assmann, Alfredo Fierro, and Leonardo Boff are available in English translation. For a critique of liberation theology, see James V. Schall, S.J., *Liberation Theology in Latin America* (San Francisco: Ignatius Press, 1982).

Chapter 6

On early Christian attitudes on war, see Jean M. Hornus, *It Is Not Lawful for Me to Fight: Early Christian Attitudes toward War, Violence, and the State*, trans. Alan Kreider and Oliver Coburn (Scottsdale, Pa.: Herald Press, 1980), Louis J. Swift, *The Early Fathers on War and Military Service*, ed. Thomas Halton (Wilmington: Glazier, 1983), and the article of Edward A. Ryan, S.J., cited in the text. For a modern Christian pacifist statement, see Gordon C. Zahn, *War, Conscience, and Dissent* (New York: Hawthorn, 1967).

William V. O'Brien, *The Conduct of Just and Limited Wars* (New York: Praeger, 1981), comprehensively explains scholastic just-war theory and applies it to recent U.S. wars. Also see Michael Walzer, *Just and Unjust Wars: A Moral Argument with Historical Illustrations* (New York: Basic Books, 1977), Paul

Ramsey, *The Just War: Force and Political Responsibility* (New York: Scribner's 1968), Joseph C. McKenna, S.J., "Ethics and War: A Catholic View," *American Political Science Review* 54 (September 1960):647–58, and *New Catholic Encyclopedia* s.v. "War, Morality of," by Richard A. McCormick, S. J.

Chapter 7

For a nuanced defense of the morality of limited nuclear war, see William V. O'Brien, "Just-War Doctrine in a Nuclear Context," *Theological Studies* 44 (June 1983):191–220. The West German and French Catholic bishops have taken a similar position. See James V. Schall, S.J., ed., *Out of Justice, Peace: Joint Pastoral Letter of the West German Bishops; Winning the Peace: Joint Pastoral Letter of the French Bishops* (San Francisco: Ignatius Press, 1984). For the different position of the American Catholic bishops, see *The Challenge of Peace: God's Promise and Our Response* (Washington, D.C.: U.S. Catholic Conference, 1983).

On technical questions, see Leslie H. Gelb, "Is the Nuclear Threat Manageable?," *New York Times Magazine*, March 4, 1984, p. 26, and Albert Carnesale et al., *Living with Nuclear Weapons/The Harvard Nuclear Study Group* (Cambridge, Mass.: Harvard University Press, 1983). For skeptical views on proposed antimissile systems, see Ashton B. Carter, *Directed Energy Missile Defense in Space* (Washington, D.C.: Government Printing Office, 1983), and McGeorge Bundy et al., "The President's Choice: Star Wars or Arms Control," *Foreign Affairs* 63 (Winter 1984–85):264–78. Freeman Dyson, *Weapons and Hope* (New York: Harper and Row, 1984), proposes ways to use technology to guide the arms race away from nuclear weapons toward conventional weapons.

Chapter 8

There is, unfortunately, little theoretical literature on the morality of military intervention in social revolutionary wars. Most writers touching the subject are more concerned with arguing particular cases than with articulating general theory. An exception is William V. O'Brien, *U.S. Military Intervention: Law and Morality* (Beverly Hills: Sage, 1979). Also see O'Brien's *War and/or Survival* (New York: Doubleday, 1969). For an analysis of the current revolutionary situation in Central America from a variety of perspectives, see Robert S. Leiken, ed., *Central America: Anatomy of Conflict* (New York: Pergamon Press, 1983).

Index